Built and Used by Poultrymen

by Big Four Poultry Journal

with an introduction by Jackson Chambers

IMPORTANT NOTE & DISCLAIMER

IMPORTANT NOTE :

As with all reprinted books of this age that are intended to perfectly reproduce the original edition, considerable pains and effort had to be undertaken to correct fading and sometimes outright damage to existing proofs of this title.

At times, this task can be quite monumental, requiring an almost total rebuilding of some pages from digital proofs of multiple copies. Despite this, imperfections still sometimes exist in the final proof and may detract slightly from the visual appearance of the text.

Some images may suffer from reduced quality due to anomalies in the original scan.

DISCLAIMER :

Due to the age of this book, some methods or practices may have been deemed unsafe or unacceptable in the interim years. In utilizing the information herein, you do so at your own risk.

We republish antiquarian books with no judgment or revisionism, solely for their historical and cultural importance, and for educational purposes.

Self Reliance Books

Get more historic titles on animal and stock breeding, gardening and old fashioned skills by visiting us at:

http://selfreliancebooks.blogspot.com/

introduction

Here at **Self-Reliance Books** we are dedicated to bringing you the best in *dusty-old-book-knowledge* – this time, an extremely useful old book to help you build essential Poultry buildings for your Sustainable or Off-Grid lifestyle.

This special edition of **Built and Used by Poultrymen** was written by the *Big Four Poultry Journal*, and first published in 1917, making it just over a century old.

The book contains sections on *Modern Breeding Houses, Modern Colony Houses, Houses for Indoor Brooder, Houses for Males or Small Pens, Adapting Barns for Poultry, Movable Fences and Fencing*, and more.

A fabulous old book and an essential read for anybody who wants to build their own old-school Poultry buildings. Especially useful for those living Off-Grid, Sustainably, or those seeking practical solutions for long-term preparedness.

~ Jackson Chambers

State of Jefferson, April 2018

Built and Used
by Poultrymen

An Attractive House for the Renter
Full Description of this House to be Found on Page 37

INTRODUCTORY

This is a revised edition of Built and Used by Poultrymen, first published by the Standard Company who sold all rights to same to the Successful Poultry Journal Publishing Co., publishers of the Big Four Poultry Journal.

We have incorporated in this book many of the practical, useful, easy-to-make houses and equipment explained in the first edition of this book as well as some illustrations and explanations of houses that have since come into vogue and have proven very satisfactory.

Every house or piece of equipment explained and illustrated herein has been built and used by some practical poultryman and found satisfactory. We are not advocating theory herein but giving the reader tried and proven houses and equipment.

If there is anything that is not understood, the Editor of Big Four Poultry Journal will be glad to explain, either by personal letter or thru the columns of that publication.

THE EDITOR.

Big Four Poultry Journal, Chicago, Illinois, April 15, 1917.

CHAPTER I.

MODERN BREEDING HOUSES

1. *Situation of House.* A poultry house should be apart from other buildings. It should face the south, and be so situated that the sun can shine on it throughout the day. It should be where water is easily obtainable, and convenient of access during the winter. If the house will have yards in front of it — and fowls are usually confined in such yards or runs — there should be a suitable space from 50 to 100 feet long immediately in front and to the south of the house for these yards.

2. *Site.* In determining the site of the house, a well-drained location must be secured, as dampness and filth around the house are the causes of many poultry diseases. It is extreme folly to build a poultry house in a hallow, or on soil that is naturally damp and cold. It is preferable to have a porous, gravelly soil that will dry quickly, and that will supply the fowls with natural grit.

3. *Shade.* Fowls require shade during the warmer months. This is the primary reason why the poultry and fruit industries can be combined to good advantage. The shade and protection of the fruit trees is beneficial to the fowls, and they in turn, protect the trees from the ravages of many species of insects. However, the house should not be placed under the trees, on account of dampness and insufficient sunlight, but from 20 to 25 feet north of the trees. The latter are then in the yards.

4. *Converting Old Buildings.* On many places, both in the country and town, are to be found old buildings that can, at small expense, be converted into practical poultry houses. If there is a southern exposure the cutting of an opening, as shown on pages 8 and 21, will generally suffice. This is the age of fresh air and these open-front houses give satisfaction, especially if they are 14 feet from front to back or from the opening to the roosts so the fowls will not roost closer than 14 feet to the opening.

5. *Stationary Houses.* The most popular class of modern breeding houses is the stationary house, built from 10 to 100 feet long. Many years ago it was customary to build poultry houses with the south front of practically all glass. This fashion remained in vogue until the early 80's, when it was completely set aside, and houses with very little glass in the front were thought ideal. The tendency of today is towards shortening the length of stationary houses.

SHED CONVERTED INTO FRESH-AIR HOUSE

These are the days of fresh air. Half the air is night air, why not use it. The above depicts how many an old lean-to shed can be converted into a modern fresh-air poultry house by cutting an opening in the front, and making a straw loft overhead. Should such a shed face the south or east the opening can be open most of the time, but if it faces north we would not recommend such a change as poultry needs sunshine as well as fresh air.

dency of today is towards shortening the length of stationary houses. Few are built over 100 feet long, and many breeders are now using houses only 8 to 15 feet long.

6. *Plans Have Been Changed.* The construction of poultry houses has been completely revolutionized in recent years. Up to 1905 it was thought necessary to board and cover with building paper poultry houses inside and out — to make them as warm and close as possible. In the following year the fresh-air style of house was advocated, and this system is now conceded to be the best,

because of the admirable results in egg production and healthy birds it insures.

7. *Outside Dimensions.* Considering the house in detail, we will commence with the outside dimensions. The length should not be over 150 feet, and the width from 12 to 15 feet. This width refers particularly to houses with many pens, as a width of from 6 to 10 feet is suitable for small portable houses. If the long house is to have a 3-foot alleyway along the north wall, the width is 15 feet.

8. *Roof.* There are two styles of roofs: (1) The shed roof; (2) the double-pitch roof. There is little difference as to the cost, and it is best to use the shed roof for houses up to 10 feet wide, and the double-pitch style for wider buildings. In the shed roof design, the height of the front studs is either 6 ft. 6 in. or 7 ft., and of the rear, 4 ft. 6 in. to 5 ft. In the double-pitch plan, the front studs are 5 ft., the center 7 ft. or 7 ft. 6 in., and the rear 4 ft. 6 in. For a house 12 ft. wide, the center studs are placed 7 ft. from the front wall.

9. *Foundation.* In building a poultry house to last, it is economical to set it on a cement, stone or brick foundation. This need not extend more than 2 ft. in the ground, and the top should be 3 in. above the ground level. Instead of the above foundation, 4-ft. posts painted thoroughly with a hot coal tar last many years.

10. *Sill.* In houses of moderate size, the sills of the frame may serve as the foundation. Use a 4 x 4-in. sill for houses under 75 ft. long and 4 x 6-in. (with the 4-in. face horizontal) for longer houses. The lumber for the sill should be painted with hot coal tar. The studs are 2 x 4 in. and the sills 4 x 4 in.

11. *Sheeting the Frame.* After the frame is erected, the studs should be covered outside with one layer of boards on the ends and back. The boards should be planed on one side, with the planed side nailed against the studs, so that it will be easier to limewash the interior. The front should consist mainly of windows and fresh-air openings. The boards should be covered on the outside with 2-ply roofing paper; the roof boards with 3-ply roofing or shingles. Don't use the shingles unless there is considerable slant to the roof.

12. *Windows and Fresh-Air Openings.* It is not necessary to use windows in the south or front wall of the house. Large or

small openings can be simply covered with 1-inch mesh wire netting, and this plan under certain conditions has given good results. We prefer, however, to have the openings smaller; to cover them with heavy muslin (cotton) in addition to the 1-inch mesh wire, and to have sufficient windows for light. The use of muslin is advisable, because it prevents drafts in the house, but all the openings cannot be covered with muslin, since it becomes dirty through use and does not let in sufficient light.

A HANDY PIECE OF MITE-PROOF EQUIPMENT

Mites are nocturnal pests. They live in the cracks and crevices of the house, coming forth at night to do their deadly work. The above equipment has many good features. It stands out away from the wall and the only way that the mites can get to the nests and roosts is to climb the table legs. This can be stopped by wrapping a wick around each leg just below the nest line and keeping this wick soaked with oil, as oil is deadly to lice and mites. The nests are on a shelf under the dropping board. The front has a hinged cover, making a dark nest, entrance being from the rear. The cover lets down easily for gathering the eggs and the nest boxes are easily removed for cleaning. The table top is not nailed down so it, too, can be easily removed for cleaning or burning if desired. The roosts rest in sockets in which are laid oil soaked cloths so that no lice or mites can get onto the roosts without crossing the oil, which is sure death. This equipment is well thought of by those who have used it, is easy to build and easily knocked down and moved.

13. *Mr. Endsley's House.* An admirable house in which the front contains windows and smaller muslin-covered openings for ventilation was built three years ago by Mr. Endsley. He says that it has given the best of satisfaction, and describes its construction thus: The house is 14 x 54 ft. The sills are two 2 x 4's placed edgewise on posts 3 in. above the ground. The plates are set the same way. A 2 x 4-in. stringer runs half way between the sill and plate of the rear wall. By using matched flooring (up and down), it is necessary to have this stringer, and especially if you want a good job of drawing your siding tight. There is no stringer in the front wall, as the windows cut out most of it. The house has five windows of 8 x 10-in. glass, and four openings 24 in. square. The openings are covered with wire netting and unbleached muslin for ventilation. The entrance door is double — the outside of wood, and the inside of wire netting covered with burlap. The wooden door is only closed in severe cold weather. This system of ventilation gives an abundance of fresh air.

14. *Location and Interior Plan.* This house is placed on ground that slopes away from it. The inside is filled with 8 in. of dry earth, and makes the inside floor higher than the outside surface. This insures a dry house. The fowls can wallow and dust in the ground floor of Mr. Endsley's house. A dry floor is one of the best vermin destroyers there is. It is divided into five pens. The droppings board is at the rear 2 ft. 6 in. above the floor. It is 4 ft. 6 in. wide and runs the full length of the building. The three roosts are raised 6 in. above the board.

15. *Dividing House Into Pens.* A poultry house over 10 ft. long is divided into a number of pens. If you intend to make a specialty of the table-egg trade, it has been learned that flocks of twenty yearling pullets make a greater profit than smaller or larger flocks. For twenty birds in a pen, divide your house into pens 12 ft. wide (running length). If you intend to mate breeding pens of standard-bred chickens, pens from 8 to 10 ft. wide are sufficient size.

16. *Partitions and Alleyway.* If you have males in the pens, it will be necessary to board the partitions for 2 ft. from the ground to prevent fighting. Above this 2-in. mesh wire netting should extend to the ceiling for Leghorns, and 4 ft. for the American breeds. The partition doors should be at the front of the house, or

the highest point, and hung on screen-door, spring hinges. An alleyway is a convenience in a long house, but not necessary in any. The partition between the alley and the pens is made the same as the partitions we have just described, except the 2 ft. of boards at the bottom is not required. There is a door from the alley into each pen, and the rest of the partition is wire netting. The partitions between the pens and the "alley" house have no doors in them, but they are boarded 2 ft. above the ground.

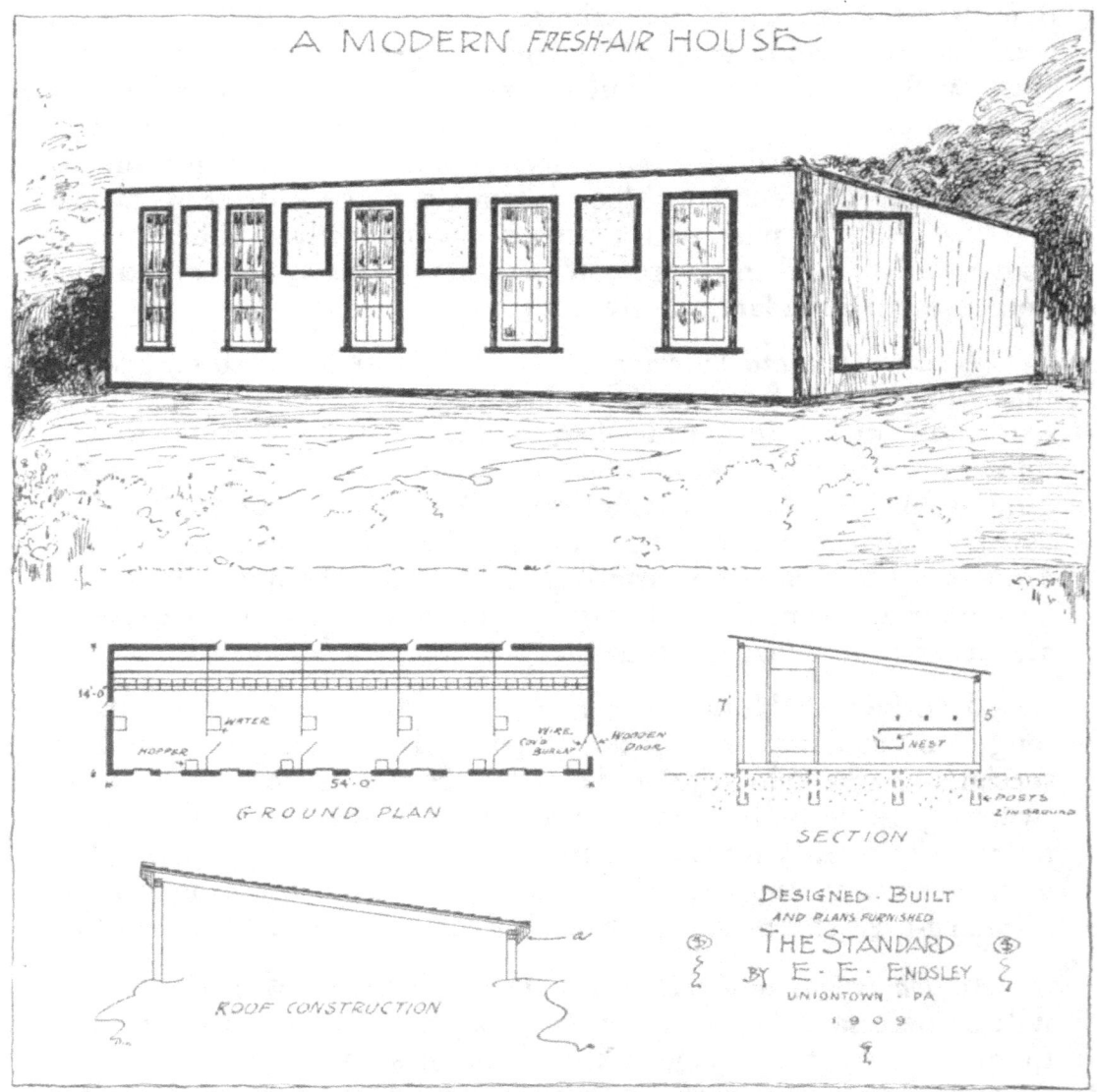

THIS HOUSE HAS BEEN THOROUGHLY TESTED FOR THREE YEARS
AND FOUND MOST SATISFACTORY

17. *Roosts.* For interior fixtures, you require roosts, droppings boards, hoppers, food and water dishes. The roost should be a 2 x 3-in. scantling, with the corners rounded off on the 2-in. edge on which the fowls roost. The top of the roost should be 10 in. above the droppings board, so the fowls can pass underneath it without having to crouch down and get soiled. It can either rest in the wire support shown in Chapter XV, or in wooden cleats nailed to the partition. It should be removable, and when there are more roosts than one, all should be on the same level. The roost nearest the wall is 10 in. from it, and the others 16 in. apart.

18. *Droppings Board.* For one roost the droppings board should be 2 ft. 8 in. wide, and for two roosts, 3 ft. wide. It is best made of ⅞-in. planed lumber to facilitate cleaning. It is usually placed from 20 in. to 30 in. above the floor, and it is preferable to have it hinged to the rear wall. It can then be raised out of the way when not required.

19. *Mr. Dick's House.* Mr. Dick prefers a breeding house 16 ft. wide and of a suitable length for the number of pens desired — each pen being 12 x 16 ft. The cross-sectional illustration gives the dimensions. The front is all open above 3 ft. from the floor, which closed part will protect the birds from winds. The open portion is covered with 1-in. mesh wire netting to keep out sparrows. The door is in the east end. It has a window in the upper half which portion is in the east end. It has a window in the upper half which swings separate from the lower half, and can be shut on windy days. By leaving the lower half of the door open, all foul air escapes. Neither part is shut, however, except when the weather is severe. Last winter Mr. Dick used a screen door covered with unbleached muslin in place of the divided door. The muslin-covered door was kept shut all the time and was all that was needed, as the air circulated through the cloth, without creating a draft.

20. *Equipment.* The roosts and droppings board are at the rear, and a burlap curtain is hung so that it can be dropped in front of the roosts on zero nights. It comes to within 6 in. of the board. The nests slide under the droppings board and are all trap nests. All the floor space is utilized by the fowls, except sufficient room for a dust box. Mr. Dick states that he would no more think of being without a dust box, than he would of letting the fowls go

without water. It's nature's way of cleaning and keeping the birds free from vermin.

21. *The Roost Curtain.* The burlap curtain which Mr. Dick drops down in front of the birds on cold nights is extensively used by poultrymen, and it more than pays for its cost. It is not in the way in the daytime; can be brought into use whenever required, and it is almost necessary when fowls with large combs are wintered in fresh-air houses. A most simple plan of arranging the curtain is shown. The supporting wire is about ⅛ in. in diameter with a round hook at each end. Two round-headed screws fasten it at

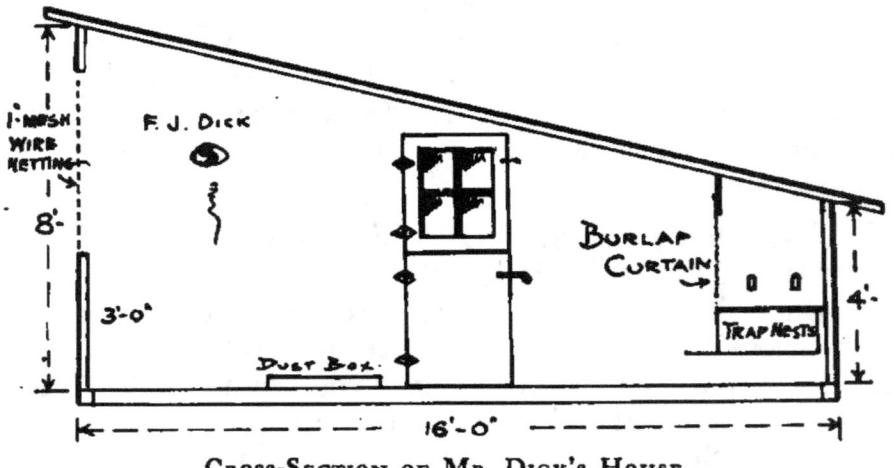

CROSS-SECTION OF MR. DICK'S HOUSE

each side of the roosting compartment. The burlap has a hem at the top, and slides back and forth on the wire like a light cloth curtain in front of a book shelf. This plan of fixing the curtain is simpler and cheaper than tacking it to a frame, or having it drop down from above.

22. *Changes Suggested.* We would like Mr. Dick's house better if it were not so wide. While the window space is high in the front wall, the sun will rarely reach within 6 ft. of the rear wall, and this is a disadvantage. In a perfect poultry house, the sun should shine during the day on all parts of the floor, so as to dry and help purify the litter. The closer we can approach this ideal condition, the more successful will our plan of house become. Keep the windows low. The lower edge of the sash should not be over 18 in. from the ground, so that the sunshine will fall on the front

half of the house almost all day. This is where the fowls like to exercise and dust. Hinged or sliding windows are advisable, as they can be opened for ventilation or for removing the litter.

23. *Low Windows Are Best.* Another advantage of the low window is that it does away with the necessity of small entrances for the birds. These little 1-foot square holes with either sliding or hinged doors and short pieces of studding necessitate considerable time in the making that could be used to better advantage. A window sill 18 in. above the ground (and a window that opens easily), is not only better than the "small hole" in many respects, but it is more satisfactory to the birds. We use the windows in this way in our houses throughout the year, and the birds do not give us any trouble.

24. *Earth Floor.* The best floor for a stationary laying house is earth. Many years ago when wooden floors were popular, we built houses with them. It is necessary to have not less than 2 in. of earth on top of the wooden floor to give the birds a pleasant ground to scratch in, and as it is almost impossible to keep this earth perfectly dry, the boards are soon rotted. Another trouble that we experienced continually was the waste of grain and the loss of chicks by rats. Although we owned a first-class Fox Terrier dog at the time, and fought the rats continually, yet they had a harbor of refuge under the wooden floor of the house from which we could not rout them. When the floor was removed it was completely rotted, and in an unsanitary condition.

25. *Making the Floor.* The level of the earth floor should be at least 6 in. above the outside ground. It is best made as follows: Large, flat stones, or coarse coal ashes are first scattered on the earth inside the house until it is completely covered. On the coarse material place 2 or 3 in. of finer ashes or gravel. Then cover all with 3 in. of porous garden soil. Over the soil have several inches of straw or other litter.

26. *Concrete Poultry Houses.* Several readers of THE STANDARD have built and used concrete poultry houses and found them quite suitable. Mr. Dornan made a house 50 ft. long and 12 ft. wide, inside measurement, of cement blocks. His house is 9 ft. high in front by 7 ft. at the back. It faces the south; has seven windows and a door in front; one window in the west end, and a half-sash

window in the east end — which he left boarded, so he can add on to the building when necessary. The window in the east end Mr. Dornan put low down "just above the sill and arranged to slide, making a convenient door for the hens kept in this pen." The upper half of the windows are covered with a cheap grade of unbleached muslin. Openings the size of half a block (8 x 8 in.) were left in the rear of the building near the ground line, making outlets for the three other pens. Mr. Dornan's floor is "mother

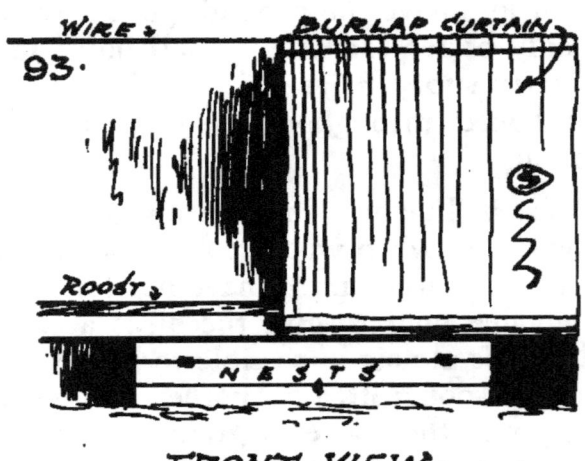

FRONT VIEW.

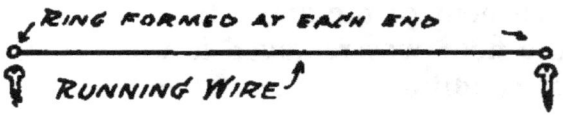

How to Make a Suitable Roost Curtain

earth" and he throws a few shovelfuls of the earth from the floor up on the droppings board, and does not have any trouble in keeping the house smelling fresh and clean. Good hemlock lumber was used for roof boards, laid on 2 x 6-in. rafters on plate of same size — the plate being golted to the blocks. The roof is covered with 3-ply tarred felt roofing.

27. *Cost.* This poultryman had no patent block machine, only a plain wooden mold that any handy man could make. He made the blocks 16 x 8-in. face, and 8 in. thick — hollow, of course. The builder says that the secret of good, hard, durable blocks is to keep them wet for at least three days. A grout wall was made 16 in.

below the ground surface. There are about 600 blocks in this poultry house, made in the usual "5 to 1" proportion. The cost was: $60 for hired help in drawing gravel, laying up blocks, putting on roof, etc.; $25 for lumber and $30 for cement, making a total of $115. However, this would not cover the usual cost of construction if one had all the work done by hired labor, as Mr. Dornan is by trade a stone mason and made the blocks himself.

28. *Stucco Houses.* Concrete poultry houses can also be built of stucco-work. Metal lath is nailed on the studs, and a scratch coat ½ inch thick is applied, and pressed partly through the openings in the metal, roughing the surface with a stick or trowel. The scratch coat is composed of 5 parts Portland cement, 12 parts clean, coarse sand, 3 parts lime and a small quantity of hair. When the scratch coat has set well, a finishing coat is applied ½ inch to 1 inch thick. This coat is composed of 1 part Portland cement, 3 parts clean, coarse sand, and 1 part slaked lime paste. Unless the house is extra long, two coats are not necessary. The finishing coat alone placed over the metal lath will give excellent results. The finish can be smooth — made so with a wooden float — or rough by rubbing with burlap. These houses could have concrete floors and are then practically vermin and rat-proof.

29. *Block Houses in Nebraska.* Some poultrymen think that a house built of concrete blocks is damp, and while several have proven by actual experience that this is not so, a stucco-built poultry house will certainly be as dry inside as any building made of wood — and more durable. Mr. Nelson of Nebraska has a number of 14 x 50-ft., cement-block houses to accommodate from 500 to 700 Leghorns. He states: "I get an 80 per cent egg yield the year around from my Single-Comb Brown Leghorns that are kept in these houses." He adds that they are never damp and that the floor is of dirt — filled 6 in. above the outside ground. He strongly advocates their use on account of their resistance to climatic changes. Any plan of poultry house in this work can be built of cement blocks or of stucco, as well as of lumber.

30. *Valuable Conclusions.* Prof. Dryden of the Oregon Experiment Station has thoroughly studied the question of poultry house construction, and we will present some of his conclusions. He believes that poultrymen have made a mistake in the past in

not recognizing the fact that the housing of poultry is an artificial condition — for our domestic fowls still retain considerable of the wild nature. This is shown when they persist in flying into the trees to roost in preference to seeking the shelter of the houses provided for them.

31. *Evil Results from Large Windows.* It has never been demonstrated that fowls can be kept in warm houses without injury to the constitutional vigor of the flock, and the vigor of his birds must be the first concern of the poultry breeder. The experiments at some of the Stations indicated that it is possible to get a slightly better egg yield by artificially heating the house, but the increased yield was at the expense of the fowls' vitality. In a warmly built house with large windows in the south, the temperature will be high when the sun shines, and at night it will be very low — the glass permitting the escape of heat as readily as it entered. During the day, the warm air of the house is accumulating moisture, and at night, owing to the great fall in temperature the air becomes very humid, often totally saturated, as is seen when moisture condenses on the walls. However, it is not dampness on the walls that is bad, but it is the moisture in the air, and the dampness on the walls is a sure indication that the air is as damp, or humid, as it is possible to be. Chickens would be better roosting in the trees, than in a house where such conditions prevail.

32. *Improving the Conditions.* By taking out the windows altogether, these conditions are improved. They would be still further improved by cutting out a larger portion of the front, and this would make what is generally known as an "open-front" or "fresh-air" house. The open-front idea may be adapted to a small house, or to a large one, a colony house or a stationary house. The open-front house has been tried in all sections of the United States, in the cold as well as in the warm. In cold sections some poultrymen cover the opening with a curtain of ordinary muslin, and when a large section of the front is thus covered, it is called a "curtain-front" house. As the material is thin there are no drafts, and the temperature of the house is nearly that of outdoors. Prof. Rice of Cornell University found that there was only one and one-half degrees higher temperature for the inside of a curtain-front house than for a similar house with a wire front.

33. *Advice.* Prof. Dryden is discussing the merits of a different style of house than any we have referred to. A curtain-front house is one in which the entire front is wire netting covered inside with unbleached muslin. We would not advise poultrymen in a cold climate to build such a house until they have had experience in the business and have kept their fowls through several winters. The amateur poultryman should by all means build a house like Mr. Endsley's with windows and muslin-covered ventilators in front. Except for a southern climate with its warm winters, Mr. Endley's house will give the fowls all the fresh air they require. In the summer the windows can be removed if thought advisable.

CHAPTER II.

MODERN COLONY HOUSES

34. *The Terms Portable, Movable and Colony.* This chapter will include the construction details and plans of all types of small houses. Poultrymen use a number of terms to refer to these houses — some call them portable, others movable or colony. The terms are synonymous, in so far as the poultryman is concerned. The colony system of housing poultry is recommended for adoption wherever possible. It consists of keeping the fowls in small flocks, and providing suitable houses that can be moved to any portion of the plant. Where the colony system of housing is used on farms, more fowls can be kept as layers and more chicks reared annually, than where the mature and young birds run in one large flock around the farm yard door. Farm fowls kept in colony houses are healthier and less inclined to attacks of contagious diseases; they require less food; lay more eggs; destroy injurious insects, and, if kept in large flocks, fertilize the land.

35. *Value of Portable Houses.* Every farmer or fruit grower who keeps poultry is strongly advised to built portable houses, and to maintain flocks of poultry in the fields and orchards. During the summer the houses are placed on pasture land; in autumn, when the grain is harvested, they are removed to the stubble to pick up any fallen grain. Here, the laying fowls and larger chicks will retain vigorous health and grow a rugged constitution with practically no attention, except watering. If a stream or natural water supply is convenient, the feed and water problems are solved. The flocks should not be crowded, or contain a great number of birds. Large flocks are invariably less satisfactory, and they will yield smaller profits. If one house is placed in a corner of the field, another near the shade of a clump of trees, etc., fences are unnecessary, as the birds quickly learn where their quarters are. No portable house should remain long in one place, or the advantages of the colony system are annulled. When the ground around the house becomes soiled, the latter should be hauled a few yards away.

36. *Points on Construction.* A portable poultry house should be strongly made, so as to withstand the strain of being hauled from one place to another. It should be well ventilated, while at the same time free from drafts, and built so it can be readily cleaned. The number of fowls the house will accommodate depends, to some extent, upon the ventilation provided. In a well ventilated house 1½ sq. ft. of floor space is sufficient during the

OTHER CONVERTED FRESH-AIR HOUSES

Above are depicted two other old sheds that were turned into satisfactory fresh-air poultry houses by opening up the south side or end and fitting these openings with a muslin covered frame so that they can be closed when the wind beats in on them. Many an old building can be converted by this method and with very little expense made into attractive and comfortable houses.

summer for each adult bird; in the winter a more liberal space allowance is advisable. Wooden floors are not necessary, but the chief objection to them is their cost. Personally, we would not build a portable house without a wooden floor, as it makes the bottom of the house tight and warm, and prevents drafts blowing on the fowls from below. A wooden floor is an advantage, also, in that you can close the house and move all the birds — confined inside — to the new location. Without a tight wooden floor in the house, this could not be done, as the fowls' legs would get caught under the sill and broken.

37. *Three Portable Houses.* A practical type of portable house is used by Mr. Brown of Oregon. This house is 6 x 12 ft. in size, 3 ft. high in front and 5 ft. at the rear. The peak of the roof is 8 ft. high. This plan of double-pitch roof house is very popular in

the eastern states. Mr. Brown has a number of them, and he moves them between the rows of hops. They have wooden floors. The Krebs' house is built along similar lines, and is also used on a

AN OPEN-FRONT COLONY HOUSE ON THE KREBS' FARM

hop farm. The Utah house was designed and tested at the Utah Experiment Station, and gave admirable results. Instead of the wire front, a muslin curtain is added for protection in windy weather.

38. *Inexpensive Houses from Old Style House.* In Paragraph 5 we told of an old style poultry house 800 ft. long being torn down and the lumber used to build modern houses. An illustration is shown here of the inexpensive houses Mr. Miller built from the old lumber. The dimensions of the new houses are 15 x 20 ft., and the openings in the front are each 3 x 6 ft. Immediately under the openings you will notice a small slanting roof. This is to keep the rain from the food hopper below.

39. *Colorado Houses.* Colony houses are used at the Fairmont Poultry Farm. They are located among the peach trees, and are of the following dimensions Length, 8 ft.; width, 5 ft.; height in

The Utah House with Muslin Curtain

front, 5 ft. 6 in.; height at back, 4 ft. 6 in. Mr. Smith, the owner, has a scratching shed communicating with each house. The shed is 10 ft. long, 5 ft. wide, 3 ft. high in front, and 2 ft. at the rear. Here in the morning, the hens scratch out their breakfast of wheat, barley and cracked corn. The addition of a moderate size scratching shed is a good idea, and especially when a number of birds are kept in each flock. Mr. Smith has each building in a run 300 yards long by 10 yards wide.

40. *The Best Canadian House.* Mr. Elford of a Canadian Agricultural College says that an 8 x 12-ft. movable colony house has given him the best satisfaction during a two years' test. The house is 8 x 12 ft. in size, built on two runners, so that a team of horses

can move it from place to place with ease. The runners are 6 x 8 in. and 14 ft. long. This allows 12 in. at each end for hitching a hook, or clevis. The runners are 3 ft. apart; upon these 2 x 4-in. scantlings are placed 2 ft. apart, as sleepers for the ⅞-in. matched floor. The studding is cut 6 ft. 6 in., which allows 4 in. for the support of the ceiling, and leaves 6 ft. 2 in. head room. The siding

MR. MILLER'S INEXPENSIVE HOUSE

This poultryman tore down an old-style house 800 feet long, and used the lumber to build a number of the houses shown above.

is single-board matched lumber. The roof is double-pitch, rough boarded and covered with roofing paper. In the west gable there is a small door for ventilation, and for putting straw into the attic. The ceiling is made of narrow boards placed 1 in. apart, and the attic is filled with straw.

41. *A Large Breeding Pen.* The illustration shows the door in the end; this is quite satisfactory for summer use, but if it is desired to haul the houses close together in the winter, a door in the side is more convenient. In that case, the window and door would be interchanged. The floor space of Mr. Elford's house is

96 sq. ft., and 25 hens and three cocks are kept in it during the breeding season. The eggs from these birds have been quite fertile.

42. *Twenty Below Zero Inside.* During the winter the temperature goes down to 35 and 40 degrees below zero at times, as the poultry plant is exposed to prevailing winds. Mr. Elford adds

MR. ELFORD'S PORTABLE COLONY HOUSE

The above house was given a severe test for two winters in a temperature of 35 to 40 degrees below zero, and gave the best of satisfaction.

that the house is by no means warm — last winter the temperature dropped to 20 degrees below zero inside, yet none of the hens had frosted combs (American breeds.) The reason for this is that the hens were healthy, and the house dry. A curtain is let down in front of the two roosts to within 18 in. of the floor on cold nights. The straw loft absorbs the moisture, and the window is open almost every day throughout the winter. No droppings boards are used. The roosts are hinged to the wall and a chain from the ceiling supports the outer side; during the day the roosts are closed up against the wall. The house is cleaned once a fortnight in winter, and once a week in summer.

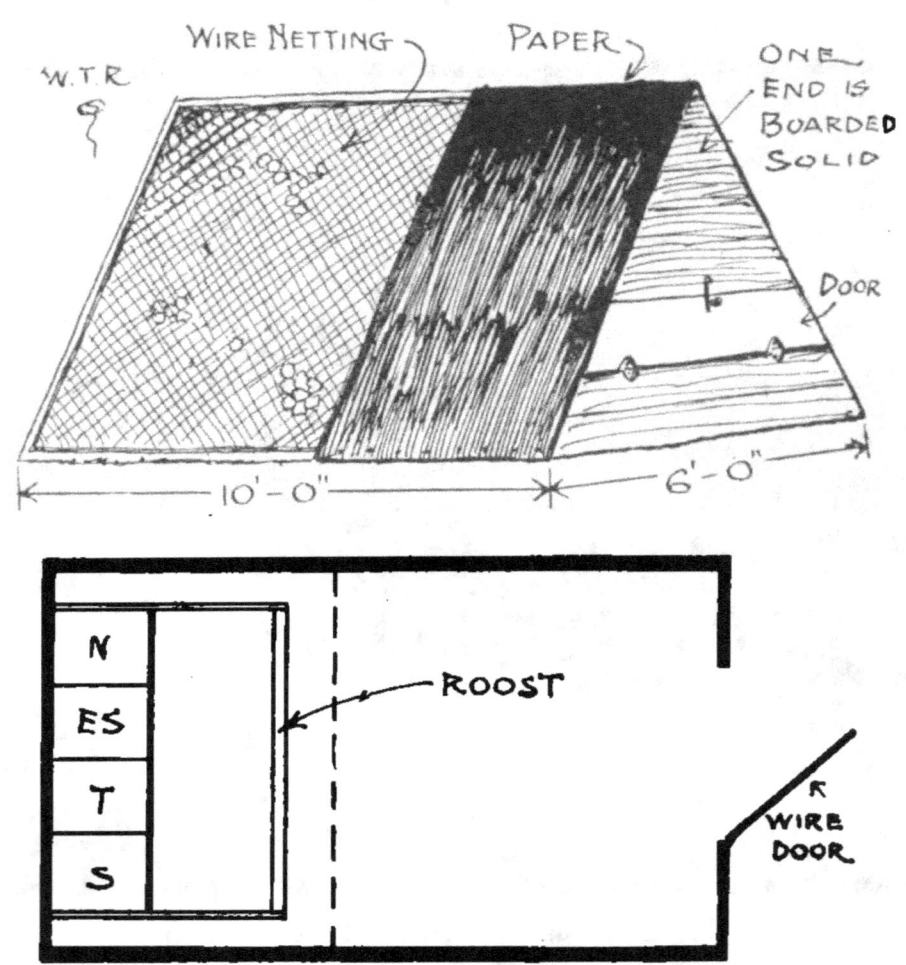

A House Costing 25 Cents a Fowl

This inexpensive but practical house was built by Mr. Roberts in Massachusetts.

43. *Method of Feeding.* Probably it would be well to say a few words about Mr. Elford's method of feeding. During the winter months there is a constant supply of dry wheat bran in a hopper — this is shown in front of the door in the illustration. A small hopper contains grit, oyster shells and beef scraps. Between two and three o'clock each afternoon mixed grain is fed in the

litter. Roots of some kind are fed regularly. In the summer the houses are hauled into the fields, where the twenty-eight fowls of each house have a yard of several acres. At this time the bran is taken away and grain is put into the hopper. The feed is brought around in a cart once a week, which, in the summer months, is the only feeding done.

44. *Houses From Dry-Goods Boxes.* It is not necessary to have a carpenter build a special colony house, as large packing boxes and a few hours' work, and roofing paper, will make a small

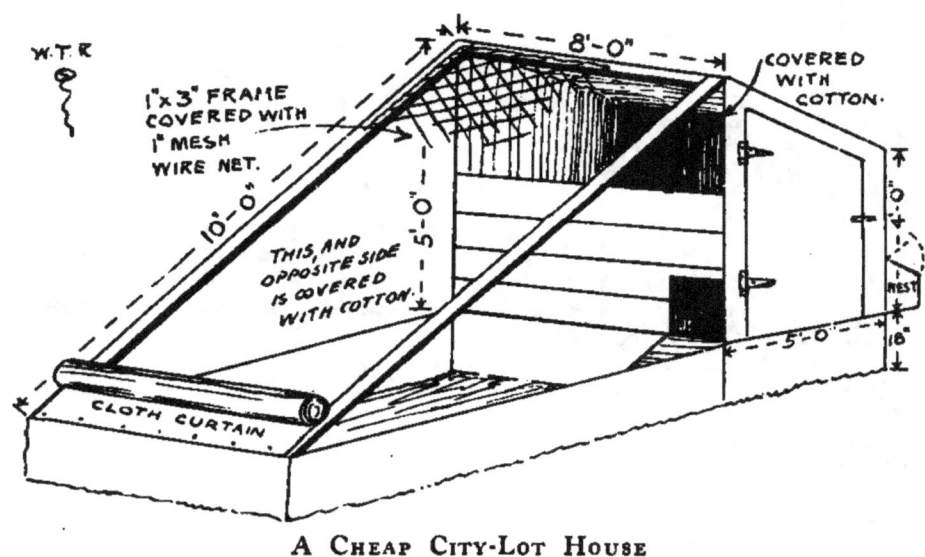

A CHEAP CITY-LOT HOUSE

Mr. Roberts built this house of packing boxes. The wooden floor is 18 inches above the ground. Note the inexpensive cotton (muslin) covered run.

house that will be quite suitable for a breeding pen, or flock of chicks. Mr. Pease made such houses of old lumber from dry-goods boxes and found them both satisfactory and inexpensive. He uses them for rearing chicks, and has a brooder heater in each. A fireless brooder would be admirable, and would be less work to attend to.

45. *A House for 25 Cents a Fowl.* Mr. Roberts of Massachusetts has designed about the cheapest poultry house we have seen. It costs less than 25 cents a fowl, and can be built with very little labor. It will accommodate from ten to twelve mature birds, and the owner uses it both in the summer and winter time. In constructing it, two frames of $\frac{7}{8}$ x 3-in. lumber 10 ft. long and 6 ft.

wide are first made. These frames are covered with 1-in. mesh wire netting to keep out the sparrows and smaller animals. They are fastened together A shape, 6 ft. apart at the bottom. One end of the house is then boarded up to form the back of the roosting compartment. The other end is covered with 1-in. mesh wire netting and a full-size door is made, so that the attendant can easily enter the house to feed the fowls. In the boarded end there

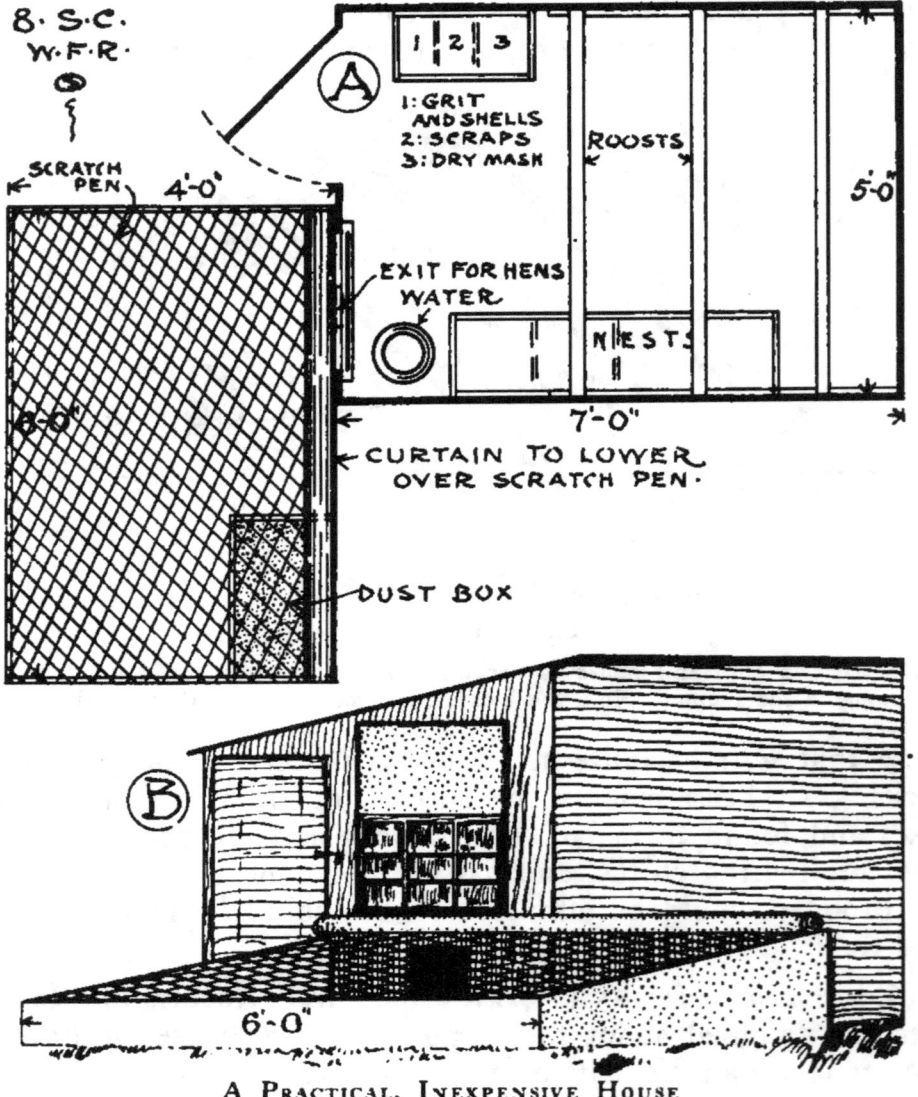

A PRACTICAL, INEXPENSIVE HOUSE

(a) Ground plan. (b) Elevation. This house was made from packing boxes by Mr. Roberts and covered with roofing paper. Nineteen hens laid well in it last winter.

is a large board on hinges, out of which the litter, etc., is thrown. The nest boxes are made from two orange cases, fastened to the boarded end 3 in. above the ground. A piece of $\frac{7}{8}$ x 3-in. stuff is nailed on each side of the nest box to support the ends of the roost.

HOUSE A, SHOWING ENTRANCE — AN ENGLISH PORTABLE HOUSE
View of a common style of an English house built on wheels.

46. *Arranging it for Winter.* Four feet of the roosting compartment is covered with roofing paper, and this makes a house and exercising pen all in one. It can be moved from place to place, so that the fowls can be kept on clean grass all summer. Before

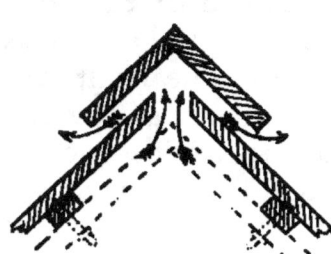

AN ENGLISH PORTABLE HOUSE
Section of roof showing the ventilation at the ridge.

winter sets in, Mr. Roberts makes a frame of 2 x 6-in. plank (set edgeways) on the ground. The frame is the same size as the ground dimensions of the wire frames, and the house is then set up on this plank foundation. A floor is built under the roost, and the north side boarded up. (The house faces east and west.) On the south side, Mr. Roberts has a cloth curtain that drops down over the wire at night, or in stormy weather. He has learned by experience that by keeping a foot of straw in the coop, the fowls will be happy and contented and lay well. We can recommend this house to poultrymen who wish to keep a few chickens in the back of their lot and do not

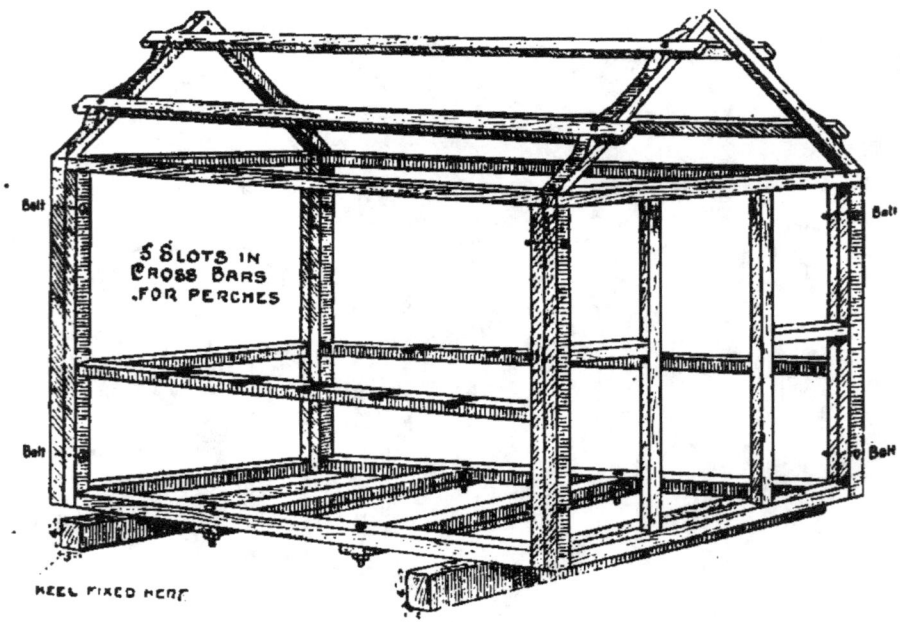

AN ENGLISH PORTABLE HOUSE
The details of the frame.

care to spend any more money on buildings than is necessary. It is also a light house that can be easily moved from one part of the city to another, when necessary.

47. *Another Cheap House.* Mr. Roberts also made and used another inexpensive house which we illustrate. It was made of two dry-goods boxes each 5 ft. by 4 ft. by 4 ft. One end of each box was removed and the two were fastened together. The house was built up to 5 ft. at the front and 4 ft. at the back. The front was boarded a little more than half way — the remainder was covered with muslin. The entire house was then covered with roofing paper, and set up on posts 18 in. above the ground. One end was made into a door. The house standing so high, made the work of cleaning and placing litter in it very simple.

48. *Cheap Houses Bring Returns.* For winter use a lean-to scratching shed was made of ⅞ x 3-in. stuff. The frame was 10 ft.

long by 8 ft. wide, and was covered with 1-in. mesh wire netting. It rested against the front as shown in the illustration. The sides of the run were covered with heavy muslin (oiled with boiled oil), and one side was a removable curtain. The top was covered with

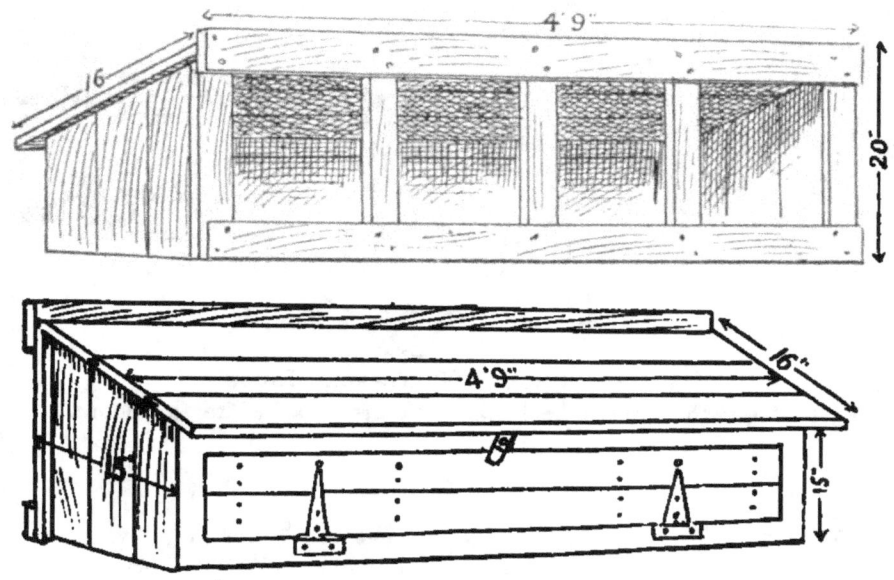

AN ENGLISH PORTABLE HOUSE
Two views of the nest box.

a muslin curtin, also oiled, that could be rolled down in pleasant weather and raised at night and on stormy days. In this lot sixteen hens were kept for three years at an annual profit of $1.25 per hen.

49. *House With a Low Run.* Mr. Roberts describes a third inexpensive house that he built and used. He is a believer in the fresh air plan, and always allows his birds an abundance of pure air to breathe. Last winter he kept nineteen hens and a male in a poultry house made of boxes. This house is also illustrated. It is 7 x 5 ft. in size, covered with roofing paper, with a outside scratching pen 6 x 4 ft. wide. The pen is covered with cloth during stormy weather and nights, and the sides are made of cloth and painted. From the nineteen hens, Mr. Roberts received from 40 to 45 per cent of eggs during the cold weather. There was not a sick hen in the flock. The dimensions and details of the house are included in the illustration.

50. *An English House.* A number of illustrations are introduced herewith of an English style of portable house. Poultrymen on the other side of the water practice the colony system of housing chickens in the fields much more extensively than we do. Their

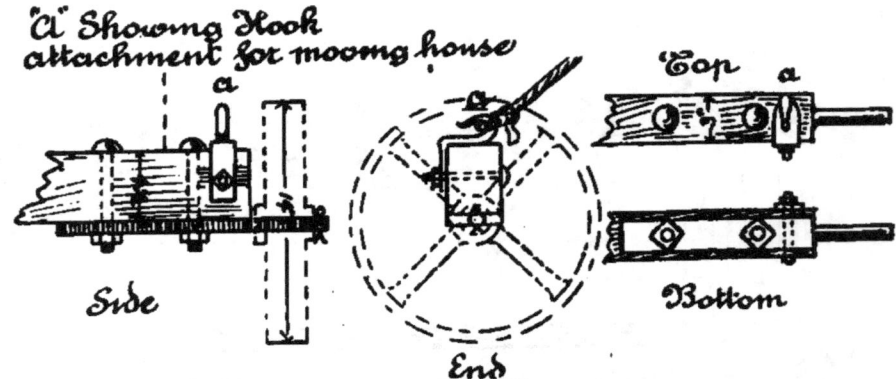

AN ENGLISH PORTABLE HOUSE
How the axle, wheel and hook are attached to the house.

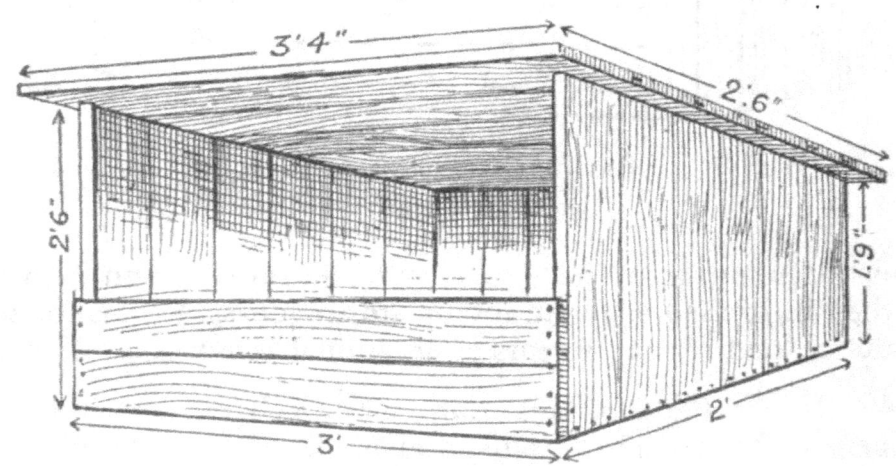

AN ENGLISH PORTABLE HOUSE
The simple dust box that is used.

houses are frequently mounted on iron wheels, and made so that they can be hauled by one or two men. Such a house is shown. The two ends are bolted to the side frames, so that the house can be taken apart, when necessary. The next box is outside the house, so that the eggs can be gathered without entering. There is a small window in one side; a door in front and a ventilator at the ridge of the roof. This is not sufficient ventilation for American conditions. In place of the small window we would cut an opening 2 x 4 ft. and cover it with muslin and wire netting. We would replace the wooden door with a frame covered with muslin and netting. This would allow fresh air to purify the house when it was used in warm weather for from twenty to thirty fowls.

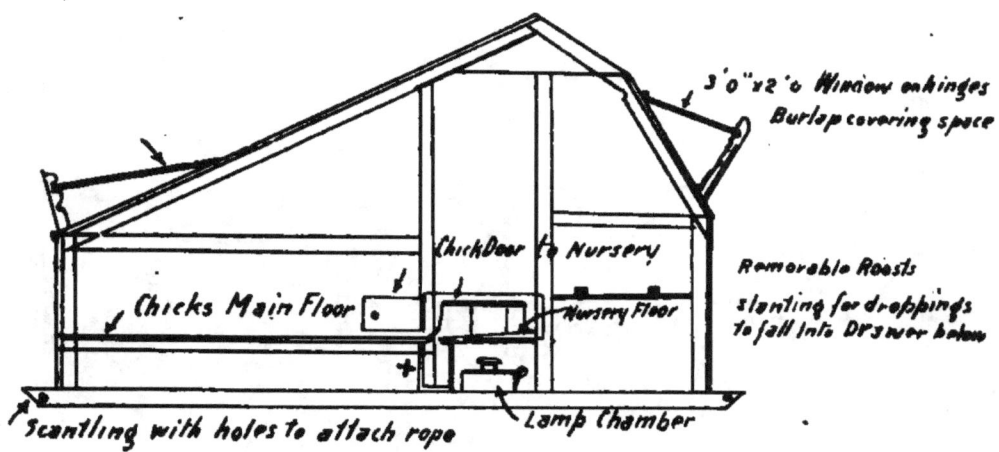

A PIANO-BOX COLONY HOUSE
The cross-section of the house that Mr. DeGraff built from two piano boxes and a few extras.

51. *Piano-Box Houses.* Colony houses of innumerable styles have been made out of piano boxes. These boxes are of a convenient size to work with, and as they can be purchased for about $2.00 each, materially reduce the cost of the house. Mr. DeGraff builds from two piano boxes, a house with about 130 sq. ft. of floor space under the roof. All you need besides the two boxes — which should be as near alike as possible — are two 2 x 4-in. scantlings 16 ft. long, one board ⅞ in. by 10 in. wide by 12 ft. long, and five ½-in. matched boards 10 in. wide by 12 ft. long, to use in filling the space between the boxes. By placing the boxes 5 ft. apart, you can make a much larger house with but little more lumber. With two pairs of hinges and a window 3 ft. by 28 in. (hotbed sash construction) you are ready to build a house.

52. *Mr. DeGraff's House.* The house is built as follows: Place the two scantlings, or runners on the ground, leveling them up for the foundation. Cut them the required length, and use the part cut off for the front connection of the raised box. Remove the back and top of the rear box and the front and bottom of the

front one. Raise the front box and lay it on its back 10 in. above the scantling, which space is closed by the extra boards.

53. *Used for Chicks.* Mr. DeGraff uses this piano-box house with an indoor brooder for raising chicks, and considers it an ideal method. The chicks having lots of room, best possible ventilation

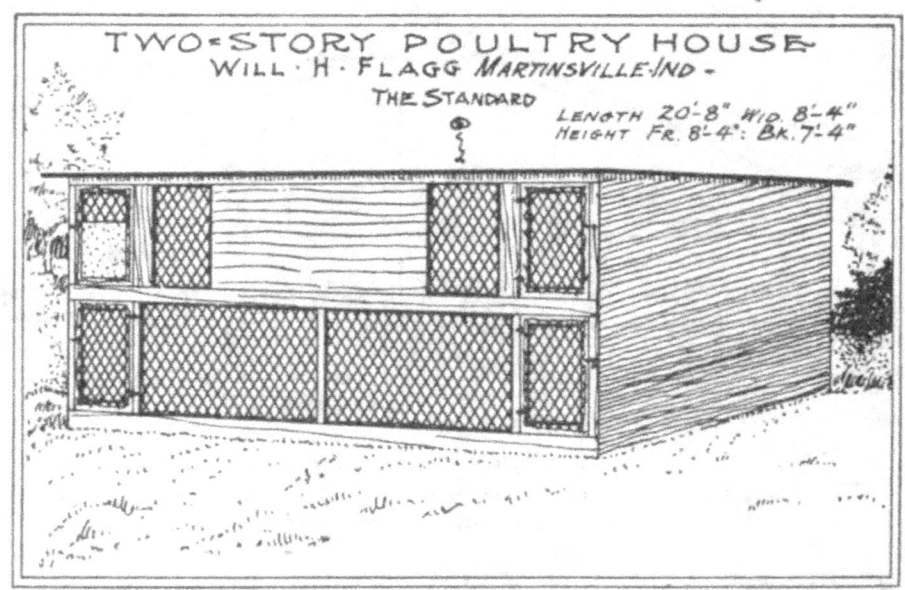

A HOUSE FOR A SMALL LOT
This two-story poultry house has two pens, and gives double floor space under the roof. Mr. Flagg has tested and recommends it.

day and night, protection from all damaging conditions, grow rapidly to maturity, and they will make either early broilers, layers or exhibition birds, as desired. Suitable brooders for this house are described later.

54. *A Breeding Pen in a Piano-Box.* It remained for Mr. Beecher of Illinois to adapt a piano box in such a way that it could house a breeding pen of Reds an entire winter. When we saw the house, it was not covered with roofing paper, and there was only a dollar or two of expense involved in its transformation. At about one-third of the height, the upper floor was laid. Six inches above the floor the roost extends from one end to the other. At one end of floor, at the back, there is a 10 x 15-in. opening (long way opposite length of box) with a 10-in. slanting board extending to the lower floor. The fowls use this slatted slanting

board in ascending or descending from their upstairs sleeping room. The nest box is at one end of the upper floor, and there is a dust box on the lower floor.

55. *How It Is Done.* Mr. Beecher has not forgotten ventilation, but has a muslin-covered opening 12 in. square at one end.

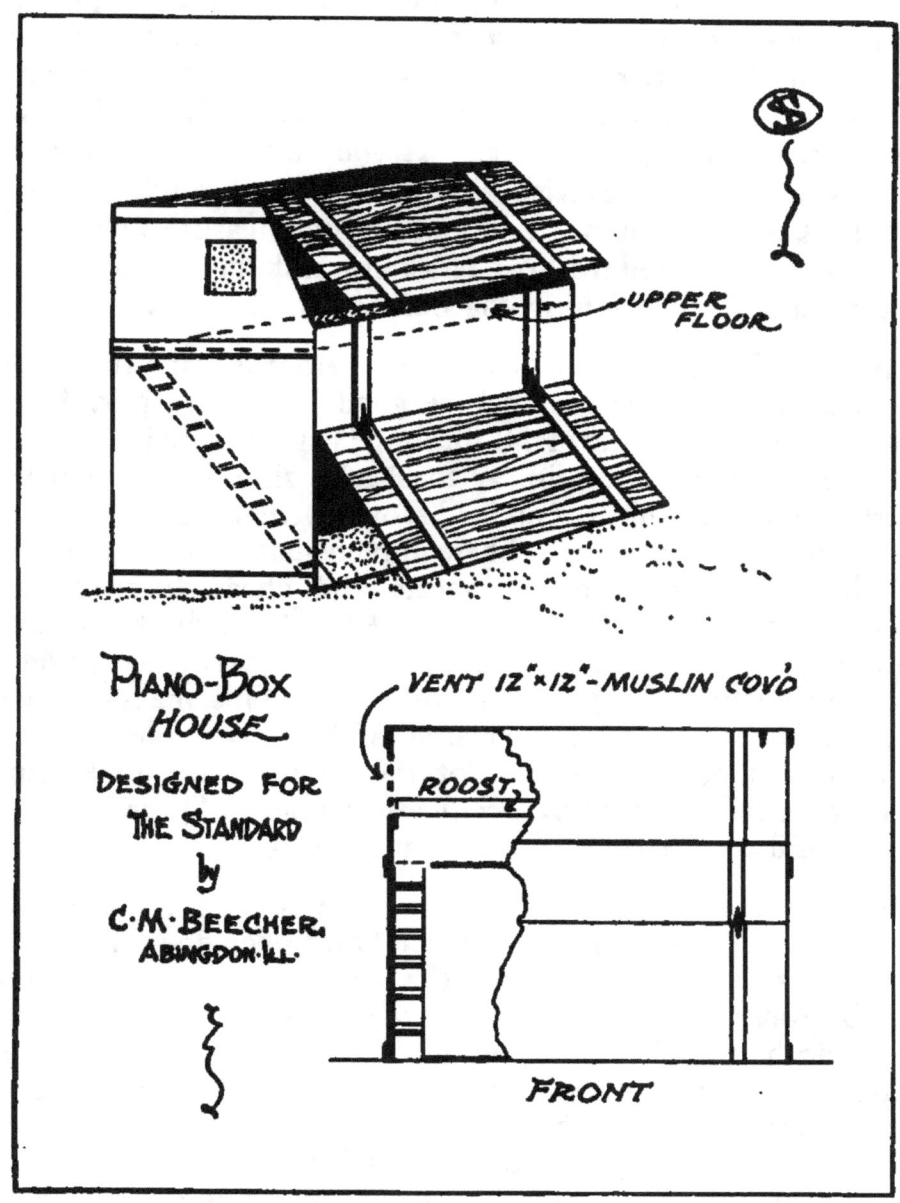

PIANO-BOX HOUSE

DESIGNED FOR THE STANDARD by C. M. BEECHER, ABINGDON·ILL·

UPPER FLOOR

VENT 12"×12"-MUSLIN COV'D

ROOST

FRONT

A NOVEL PIANO-BOX HOUSE

The latest plan of utilizing the piano-box is to make a two-story poultry house of it as Mr. Beecher has done.

There are two hinged doors in front, and in warm weather both are wide open. The position of the doors shown in the illustration, is the way they are propped open during a rain or light snow storm. It is remarkable how poultry will thrive when housed in these small, yet well ventilated houses. This pen of Rhode Island Reds (six hens and a male) laid well, and came through our past severe winter without so much as a spike frozen. In their upstairs sleeping room huddled together in a few feet of space — which they would warm by natural heat — it can be truthfully said that they were "as snug as bugs in a rug." It you have a small yard at the rear of your house and wish to keep a pen of fowls, adopt Mr. Beecher's piano-box plan. You can obtain pleasure and profit by owning a pen of standard-bred chickens, and there is a very small outlay required in beginning the business in this way.

56. *Mr. Flagg's Two-Story House.* Another two-story poultry house was built and found to give good results by Mr. Flagg of Indiana. This house is 20 ft. 8 in. long; 8 ft. 4 in. wide; 8 ft. 4 in. high in front, and 7 ft. 4 in. at the back. A solid board partition runs through the middle (up and down), making two pens each 10 ft. 4 in. by 8 ft. 4 in. These dimensions were taken because the builder had steel roofing sheets and cheap lumber that he wished to use. The house has a floor 4 ft. above the foundation — making it two-story. The lower story has an earth floor, and 1-in. mesh wire netting along the entire front. This makes two ideal scratching sheds, when kept covered with straw. There is a wire gate into each pen at the ends of the house. The upper story is reached by a ladder, kept in a convenient place. There is a door at each end and a wire-covered window in each pen besides.

57. *The Fowls Roost Upstairs.* The four doors are covered with 1-in. mesh wire netting, and the two upper doors have muslin curtains that fasten on outside. The curtain has eyes that go over buggy buttons; these eyes are an even distance apart, so that the curtain can be raised or lowered, and an opening left at the top or bottom. The fowls go up a ladder to the upper room where the nests and perches are. They roost near the partition between the upper pens, which keeps them out of all drafts. In the back wall of each scratching pen, there is a small door hung on screen-door hinges, so that when the door is open it will not blow shut in a

storm, and when shut, it will not blow open and let the fowls out. These doors lead to the yards behind the house. Mr. Flagg did not make the stories higher because he has several children that like to work among his chickens; and in fact, he does not like a high roosting pen. He built a two-story house because he has no ground to spare, and this plan gives them double floor space under one roof.

AN ATTRACTIVE HOUSE FOR THE RENTER

The illustrations on page two are almost self explanatory. This house is 6 x 9 feet on the ground, 6 feet high at the peak and 3 feet high on sides. It can, of course, be built any size desired. Back and sides are boarded with matched lumber. The roof is covered with some good roofing material, hinged at peak to allow opening, as shown in the lower illustration. Roof should have a wire covering about four inches below roofing so fowls cannot escape when roof is opened. The front is of wire as shown in illustration. Droppings board, roosts and nests are located inside at rear. In winter the space between the roofing and wire covering should be stuffed with straw or some such material and the two lower oblong panels in front filled with glass, the balance of the front being covered with muslin.

This house can be built in sections and bolted or screwed together thus making it easy to take down and move.

For a fresh-air house in a mild climate or for a summer house in any climate, we have never seen its equal.

In a house of this kind, 6 x 9 feet, one can keep a dozen fowls and get the very best of results. The builder can use his own discretion in regard to putting in a board floor.. Personally, we prefer to use the dirt floor as the fowls are kept confined and it gives them a place to scratch and the house is easily moved to get them on fresh ground.

CHAPTER III.

MODERN BROODER HOUSES

58. *Necessity of Pure Air.* Brooder houses for rearing chicks have been in use by poultrymen for a number of years. Many of them have given, and continue to give, entire satisfaction, while in others it seems to be a difficult problem to regulate the temperature and supply a sufficient amount of fresh air. We cannot emphasize the fact too strongly that all ages of poultry require pure air constantly, as it is the lack of this natural agency that has worked more havoc among the owners of large poultry plants and caused them to say "there is no money in poultry" than any other condition. We illustrate herewith an admirable hot-water heating system of moderate cost for an up-to-date brooder house, and later we will refer to a plan of one prominent poultryman in which he pipes pure air to each individual hover of chicks.

59. *Hot-Water Heating System.* The designer of the hot-water heating system is Mr. Rathbun, an expert heating engineer, and for over two years the chief draftsman in a leading firm that deals in heaters of this kind. In writing the description of the system, Mr. Rathbun stated that he would illustrate and describe the most reasonable and best brooder house heating system he was acquainted with, and one that an experienced poultryman or layman could easily install. The illustration and text refer to a house 25 ft. long, 7 ft. wide in the hover yards, and 3 ft. in the walk — making a width of 10 ft. over all — and 7 ft. high at the highest point. The walk and floor under the hovers can be constructed of concrete or plank, as may be desired, while the hover platform, which should be from 10 to 12 in. higher than the walk, can be built of boards with a 3-in. rim around to allow the filling in of the same with garden dirt for scratching. The hover platform is 20 ft. long, which gives 5 ft. of space for the setting of the boiler or tank heater. This can be set on a level with the walk, or a few steps lower in a pit as shown.

60. *A Modern Hover.* Let us suppose the hover platform to be divided into ten rooms, or yards, for the different grades, varieties and ages of chicks. Each yard should have a hover made of canvas hung under the 1-in. coils in folds, with drops or curtains on all sides. This makes it as warm as a mother hen. In using canvas in this way, Mr. Rathbun has the same method as the fireless brooder advocates, and we indorse his advice, instead of the orthodox brooder cover made of wood with a few tin pipes for ventilation. The canvas (outing flannel would be warmer) would retain sufficient heat for the chicks, and there would be a working outlet for the impure air.

61. *Double-Deck Hovers are Not Advised.* Mr. Rathbun also states that in some cases where more room is required, a double-deck hover floor can be constructed midway above. In such an event, care must be taken in installing the heating coils, and only two coils run for the first hover, so that there will not be too much heat to rise and warm the floor of the upper hover. Unless this precaution is observed, the excessive heat of the upper floor will cripple the little chicks' legs. For the second floor hover, three coils should be run, so that there will be plenty of heat to protect the chicks from the cold that is transmitted from the roof above. However, unless it was necessary to *greatly* economize space, we would not place hovers on two floors.

62. *Installation of Heating System.* This heating system will give by an even running fire, just the right amount of heat and the best of satisfaction in weather not more than 10 degrees below zero. The installation of it, together with the material required, is as follows: Place a tank heater of sufficient capacity (which for the plan as shown would be 150 gallons) and locate it where shown at the end of the brooder house.

Connect to one of the flow openings (left-hand pipe at top of heater) a 1¼-in. main and run same to hover coil manifold, consisting of three 1-in. pipes. From the manifold at the far end of the house, carry 1¼-in. return pipe back to the side opening of the heater. From the other flow opening (right-hand pipe at top of heater) a 1¼-in. connection is to be made to an auxiliary coil run around the side of the hover platform in the walk, with a 1¼-in. return from the manifold at the far end of the house back to the

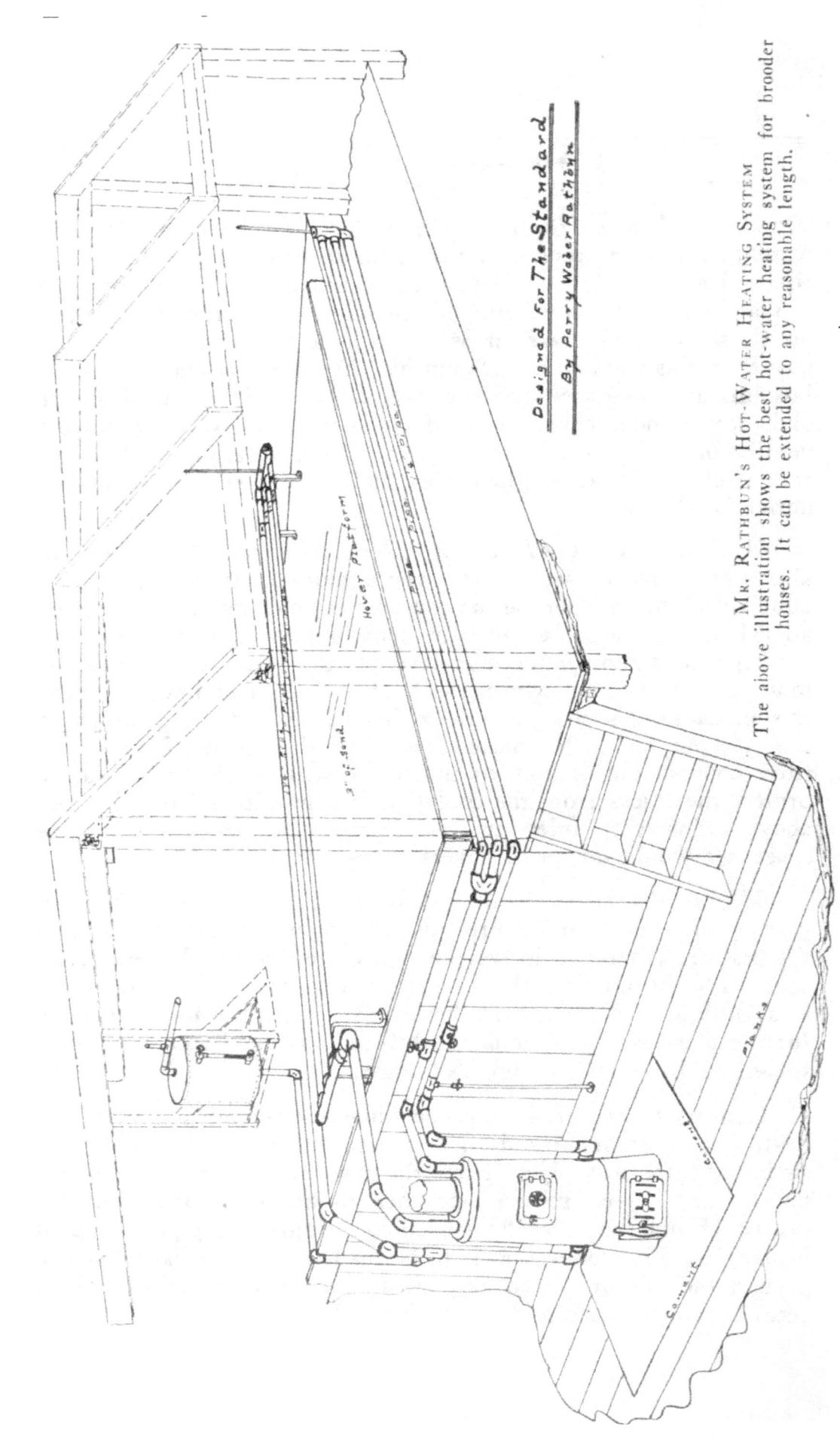

Designed for The *Standard*

By *Perry Water Rathbun*

MR. RATHBUN'S HOT-WATER HEATING SYSTEM

The above illustration shows the best hot-water heating system for brooder houses. It can be extended to any reasonable length.

side opening of the heater. On both of these auxiliary mains, valves should be placed, so that they can be put into commission whenever the temperature demands their use. A ½-in. drain pipe will have to be run from the auxiliary coil down to the floor to the sewer, so that when the valves are cut off the coil can be

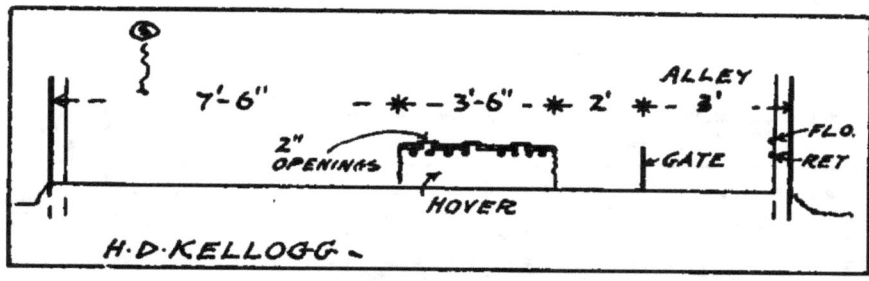

CROSS-SECTION OF MR. KELLOGG'S BROODER HOUSE

drained to prevent its freezing. From the manifold at the far end of the house, run a ⅜-in. air vent at least 3 ft. in the air, to keep this end of the heating system from becoming air bound.

Pitch all coils and mains down in the direction of the boiler at least 1 in. in every 8 ft. Take from the return at the boiler, as shown, a ⅝-in. expansion line to an expansion tank placed on a shelf, or brackets, as near the top of the building as possible, or at least 12 in. above the highest point of the system. The tank can be either of the open or closed pattern. If closed, an air vent will have to rise 2 or 3 in. above the overflow line, which is carried out to the side of the building, or down to the sewer. This heating system can be filled with a funnel attachment to the side of the expansion tank, or with city water pressure into the return at the boiler. A ½-in. drain connection should be taken from the lower part of the system to the sewer.

63. *Complete Cost.* For all the materials complete, including a 150-gallon tank heater, the cost has been estimated at the present market prices for the heating goods required at $25.60. This price will probably be exceeded later, as at the present time these materials are at their lowest cost. Those required are: Tank heater, expansion tank, water gauge for expansion tank, gate valves, drain and feed stops, manifolds, cast iron elbows, right and left couplings, assorted nipples, the necessary black pipe and a can of cement for making joints. For $2.50 additional, a hot water thermometer and

[41]

damper regulator connections to check and draft doors, can be installed. The regulator will not only mean an economy in fuel, but will keep the house at an uniform and satisfactory degree of heat.

64. *Pointers From An Engineer.* Mr. Rathbun gives these pointers. The heart of the brooding house is the heater, and this must be large enough to furnish sufficient heat to warm the chicks.

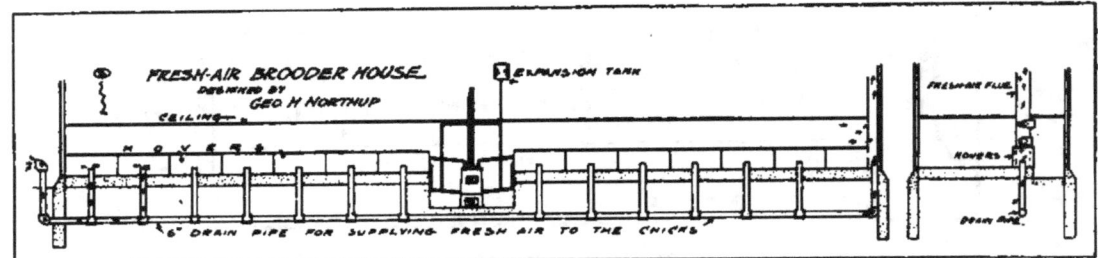

LONGITUDINAL AND CROSS-SECTIONAL VIEWS OF MR. NORTHUP'S
BROODER HOUSE

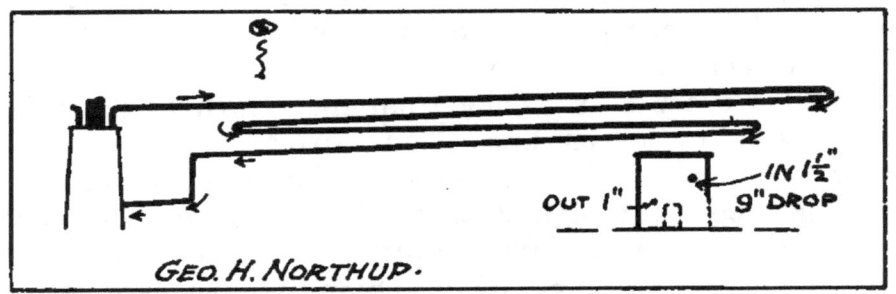

MR. NORTHUP'S HEATING SYSTEM

Notice the small cross-section of the hover showing the inflowing 1½-inch pipe
and return 1-inch pipe, also the top of the ventilating tile in the floor.

As neither too much, or too little heat should be furnished, a regulating device to automatically open and close the drafts should be considered. The several regulators on the market at the present time, are unlike thermometers that work from the temperature in the hovers only; a regulator positively opens and closes the drafts, while the thermometer is insufficient to control any fire, especially in the large heating plant required for houses from 50 to 100 ft. long. When the fire is at its height, it takes more than a check draft door to control it.

65. *Filling the House at Once.* Mr. Kellogg uses his brooder house in an original way. Instead of placing in it chicks of all ages,

he has sufficient incubators to entirely fill his brooder house with chicks from one hatching. After these are three or four weeks old, they are removed to colony houses and the brooder house is filled with the second batch. This is an admirable move, and one that we can heartily recommend. A small half-tone illustration of the brooder house is shown. The house proper is 88 ft. long and 16 ft. wide; it will house from 1,000 to 2,000 chicks comfortably. There are four 1¼-in. flow pipes and four 1¼-in. return pipes. The heater is controlled by a thermostat and electric regulator. In addition to the eight pipes under the hovers, there is a flow and return pipe on the rear wall of the building — the lower about 12 in. above the floor.

66. *The Common Hover Top.* The eight hot-water pipes are near the center of the building as shown in the cross-sectional illustration. The hover cover consists simply of a ⅞-in. door 3 ft. 6 in. wide by 4 ft. long, which rests on the pipes. From the front and back edges, two pieces of slitted blanket extend to the floor. This hover cover is the ordinary wooden top that we referred to when discussing Mr. Rathbun's canvas hover top. We prefer the latter. Mr. Rathbun's plan for the installation of the heating system is applicable to Mr. Kellogg's system, or any one that a poultryman would select.

67. *Mr. Northup's Brooder House.* Mr. Northup gives the details of a brooder house that he had used for a number of years, and of which he said: "It is the best brooder house I have ever seen or used. I have hardly any deaths with my chicks, and in fact, I have placed chicks in the house that seemed to be afflicted with white diarrhoea, and I have had them recover and do well." Mr. Northup's house is two stories high — brooder house on the ground floor and four pens above for keeping surplus stock. It is 78 ft. long, 14 ft. wide and 6 ft. high to the ceiling. It cost $800.

68. *How Fresh Air is Supplied.* The heating system is the simplest of any we have described, but the main feature, and the one to which we wish to direct your attention, is the manner of supplying fresh air to the chicks in the hovers. As is shown in the larger illustration of the house, a 6-in. drain pipe conducts fresh air from outdoors to each hover. The drain pipe is 3 ft. in the ground, and at the center of each hover, another pipe extends up-

ward from the supply pipe. The tops of the upright pipes are 6 in. above the concrete floor, and covered with ¼-in. wire netting caps, to prevent the entrance of dirt or the loss of a chick.

69. *The Interior Fresh-Air Flue.* The outside end of the supply pipe extends above the ground, and has a right-angled coupling at the top to keep out the rain. The other opening is at

GASOLINE *BROODER HOUSE*
DESIGNED FOR
THE STANDARD BY
J. M. WILEY · ELMWOOD ILL·

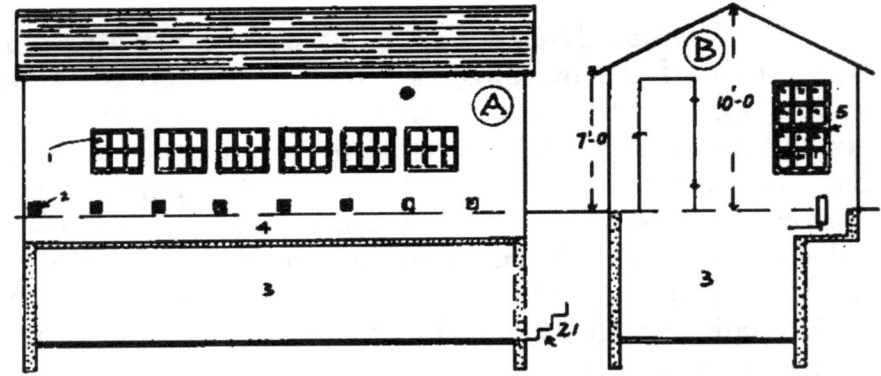

A GASOLINE-HEATED BROODER HOUSE

The letters in the illustration of Mr. Wiley's brooder house refer to the following:
A—Longitudinal section of house. B—Cross-section of house.

the far end of the house, and enters a fresh-air flue 12 x 15 in., which extends through the roof. A galvanized iron stove pipe could be used in place of the wooden flue. There are openings in the flue covered with adjustable slides near the ceiling and over the hover, so that it will act as an efficient ventilator of the house in warm weather.

70. *Heating System.* The heater is in the center of the building in a pit 8 ft. wide. Each pen for chicks is 5 ft. wide, and there are seven on each side of the heater. The smaller illustration shows the simple method of piping used. The flow pipe that leaves the top of the boiler is 1½ in., and this is reduced to 1¼, and then to 1 in. There is a drop of 9 in. in the piping. The hovers are 5 ft. long, 2 ft. wide and 2 ft. high. They are simply boxes (with

loose covers and no floors) that rest on the concrete floor, with the front of slitted cloth. The back of the box should have a hinged door, so that the chicks can be more easily examined from the 3-foot alleyway. The house rests on a concrete foundation and the entire floor is of the same material. The raised concrete floor (11 ft. wide) on which the hovers rest is 14 in. above the level of the alleyway.

71. *A Gasoline Heated House.* Mr. Wiley uses a gasoline heated brooder house that is ideal for a poultry plant of moderate size. It is 24 ft. long, 12 ft. wide, 7 ft. high to the eaves and 10 ft. to the apex of the roof. It has been in use for two years and Mr. Wiley says he is delighted with it, and that it always raises the chicks. From one to two gallons of gasoline will supply each burner one week. The gasoline tank is outside the house 3 ft. above the ground, and connected to one end of the supply pipe. The temperature of each individual hover can be easily regulated, and labor is saved by having no lamps to trim or fires to feed coal to.

72. *Two Stories.* If you examine B in the illustration, you will note that the house is built in two stories. The cellar is used for running the incubators, and the structure above is the brooder room. The rear concrete wall of the house is made with a "step" in it. This step (lamp pit) extends from one end to the other. It is 30 in. wide by 16 in. deep, inside measurements. E shows the end view of this pit with the heating arrangement in place — C and D the front view of the same.

73. *Concrete Parts.* We will consider the concrete work. When the walls and lamp pit are finished, it is necessary to divide the pit (4) by seven concrete piers (16) into eight pens 3 ft. wide. The piers are 16 in. high, 30 in. long and 4 in. thick. When the seven piers are in position, extend 2 x 12-in. sills across the house, the rear end of each sill (20) resting on the concrete pier. These sills form the foundation of the house, and they also serve as partitions between the pens of chicks.

A concrete floor (15) 4 in. thick is made between each pair of piers. This floor has a 6-in. opening in the center to fit the outside flue of the heater (7). A door of $\frac{7}{8}$-in. boards that just fits between the piers with a 6-in. piece of stove pipe at the center,

[45]

can be supported at the proper height and used to mold the concrete floor on. The concrete work is now finished and the wooden brooder house is built. There are six windows (1) in the south wall, each with six 8 x 10-in. lights. Two pens have no windows. The window (5) in the east end has twelve 8 x 10 lights — double sash.

74. *The Gasoline Heater.* The heater (7) is 6 in. in diameter and 16 in. long. Four inches of pipe extend below the concrete floor and 8 in. above. The inside flue (8) is 2 in. in diameter at the top and about 5 in. at the bottom. It conducts the gasoline smoke and vapor to the heating tank (9), where it leaves the tank as shown at (18) in E. The pure air enters between the two pipes at the bottom of the heater, and as it is warmed, rises and flows into the brooder at (19) in E. The outside pipe (7) of the heater fits smoothly into the concrete floor (15). The upper end of the inner flue (8) enters a removable heating tank (9) 10 in. wide, 30 in. long and 2 in. thick. There is a small drawing of this tank at E among the parts. D shows the heating tank and brooder cover (10) removed from the heater.

75. *Fumes Do Not Enter Hover.* You will notice in E where the gasoline fumes are escaping at (18) that the brooder cover does not extend to the rear wall. There is a $\frac{7}{8}$-in. wooden partition immediately below the back edge of the brooder cover. This construction is necessary in order to allow the fumes to escape into the house. If the brooder cover extended to the rear wall, the gasoline fumes would all enter the hover and injure and poison the chicks. At the front of the brooder cover there is the usual slitted blanket to retain the heat.

76. *Burner and Piping.* Among the parts shown in E (11) is the gasoline burner. It is a gasoline torch burner complete, and is supplied with two regulating vales. It is commonly used by fruit peddlers, "fakirs," etc., and costs 80 cents. As shown in D and E, the $\frac{1}{4}$-in. pipe (14) that brings the gasoline from the outside tank to the burners (11) runs along in front of the concrete piers. Other $\frac{1}{4}$-in. pipes extend from the supply pipe to the burners. It is necessary to have two valves on the short supply pipes, so that the burner can be lowered free from the heater and swung around to

the front, when it is necessary to clean it. When the gasoline burner is in operation, it stands up a couple of inches inside the flue (8). This explains the necessity for lowering the burner, before it can be brought around to the front for examination.

77. *Advantages of Gasoline Heater.* On account of the uncommon construction of this gasoline-heated house, and the number of details it is necessary to give, it appears somewhat more complicated than a hot-water house would be, but it is really not so. When we consider that the principal parts are built of concrete, and that the heating equipment is inexpensive, practically indestructible and placed where it can be easily reached; that it can be simply regulated and operated at small expense, then the great value of this house for the moderate size poultry plant is apparent. This is the ideal house for the average poultryman. If he wishes to economize in construction, he can omit the incubator cellar and concrete work, but this would make it necessary for him to get down on his hands and knees when he wanted to regulate the gasoline burners. This, however, is not an important objection, as the operators of most brooders understand.

CHAPTER IV.

HOUSES FOR INDOOR BROODER

78. *Special Houses for Chicks.* Although any style of colony house or building can be used as a shelter for brooders and chicks, and there are a number of such houses in the preceding chapters, we describe and illustrate two houses that have been especially designed for the purpose. The first portable house is inexpensive and simple to build. It can be used in the spring months for one or two brooders, and later on it can be hauled to a field for the larger chicks. It is built of $\frac{3}{8}$-in. lumber planed on one side with the smooth side placed against the studs. This method of construction is recommended instead of having the smooth side out, for the reason that the interior of the house can be limewashed much easier. If minimum expense is a desideratum, the house can be built of lumber taken from large dry-goods boxes. A small amount of 2-ply roofing paper will cover the outside, but not less than 3-ply roofing paper should be used for the roof.

79. *Construction of the House.* In the lower part of the illustration the frame is shown before it is covered with the siding. The two runners are made from a 2 x 4, 16 ft. long cut in two diagonally. On these runners a floor 6 x 7 ft. is laid. For the front and back edges of the floor, 2 x 4-in. sills are nailed securely to the runners. On these sills the four corner and middle-front studs rest. The studs at the right of the door and at the center of the west stand on the floor boards. On top of the six end studs pieces 2 in. square by 8 ft. long are fastened. These are made from a 2 x 4, 8 ft. long, cut in half.

80. *A Pointer on Muslin Frames.* The west half of the front is covered from top to bottom with 1-in. mesh wire netting 3 ft. wide, stapled directly to the studs. The remainder of the frame is then covered with the siding. There are three wooden strips $\frac{3}{4}$ by 2 in. by 6 ft. long covered with three strips $\frac{3}{8}$ by 4 in. by 6 ft., that form the three runners for the oiled muslin frames. These six strips are nailed at the top, middle and bottom of the front of the house, as shown in the upper illustration. The sliding frames are then made (see the small drawing of one frame) and covered with muslin. This is painted with boiled oil so as to render it waterproof. It can be stated here that the muslin-covered frames used in ventilating poultry houses are not treated with boiled oil, or any other paint. When the muslin is oiled, it not only renders it waterproof, but practically air-tight. In this brooder house, the frames can be air-tight because they are always kept slightly open for ventilation.

81. *Object of House.* The main object of a house of this kind is to have a bright, well ventilated, dry and safe pen for little chicks. Any indoor brooder can be operated inside, and by standing a 10-in. board on edge across the middle of the house, with the brooder near the wire front, the little chicks can be kept in the sunlight during the first few days. The floor of the house should be covered with clover chaff, and prepared chick food, cracked wheat, millet seed, oatmeal, etc., should be thrown in the chaff several times daily. When the chicks are one week old they can be given the entire house to exercise and feed in, and allowed to enter and leave the brooder at their own pleasure.

82. *Further Good Features.* Other advantages of having the oiled muslin frames in front are that the wire openings can be closed when a driving rain is beating against the house; that either the upper or the lower frame can be closed while the other is open, and that the frames can be partially open at night to supply sufficient ventilation. The slanting board in front of the chick door is hinged and should be raised and fastened at night to prevent the entrance of rodents. With this door closed, the 1-in. mesh wire front effectually keeps out any animals that would harm the chicks.

The poultryman should build several portable houses of this kind. They will easily repay their cost in the saving of chicks in cold, rainy weather. While the chicks in coops outdoors will be cold, wet and insufficiently nourished, the colony-house chick will be happy and contented from morning to night, searching for small grains among the litter. Besides, it is a great convenience to a poultryman to be able to quietly look after his chicks and brooder

A COLONY HOUSE FOR CHICKS

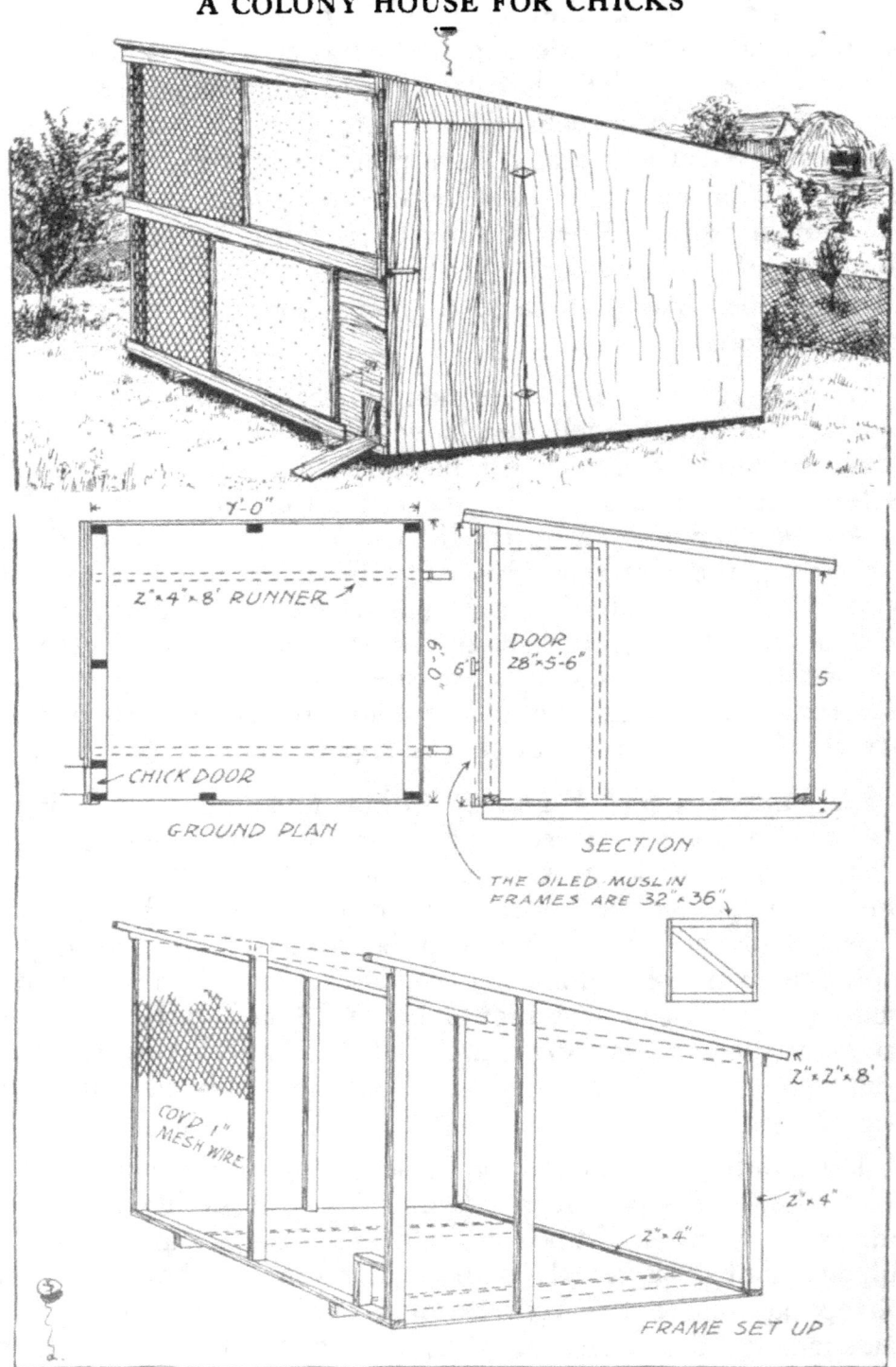

A PRACTICAL, INEXPENSIVE HOUSE FOR SHELTERING BROODERS

inside a dry house in inclement weather. A fireless brooder in a house of this kind would make an ideal outfit for raising fifty chicks to maturity.

83. *A Colony House Brooder.* Mr. Harris' colony house is 5 ft. long, 4 ft. wide, 4 ft. high in front and 3 ft. high at the rear. The entire front is glass so as to admit lots of sunshine. This is a great germ destroyer and is essential on cold, wet days. During the summer time it would be well to stand boards in front of the greater part of the windows, to prevent the interior of the house becoming too hot. In the end facing the east are the doors. The outside door is wood; the inside is a screen door made of 1-in. mesh poultry netting. When the wind is not against that end of the house, the wooden door is left open, when it is, the wooden door is shut and the window behind the door opened. The latter is covered with unbleached muslin and makes a good ventilator without any draft, so long as the wind is against it.

84. *Mr. Harris' Brooder.* The brooder is 4 ft. long, 2 ft. 6 in. wide and there is a partition in the middle. This makes a double brooder or a single brooder, as Mr. Harris used it the first year. Each compartment has a hinged window in front; the roof is also hinged, so that it is easy to get at the brooder to clean it. The hover is 2 ft. square. It is made of $\frac{1}{2}$-in. pine boards, and the under side is covered with flannel hanging loose (like a fireless brooder), and the front is of flannel cut and fringed. The heater is 12 in. in diameter; placed in the middle of the hover to prevent the chicks from crowding into the corners, as they are very apt to do. It is made entirely of sheet iron. In the center of the bottom is a hole 3 in. in diameter, and this is where the lamp chimney is placed. Two inches above the bottom (horizontal) there is a piece of sheet iron 11 in. in diameter. It is supported by wire rods, and is used to spread the heat and make the heater more efficient. The fumes of the lamp escape above the center sheet by the $\frac{1}{2}$-in. space around the edge. In the center of the top of the heater there is a pipe 1 in. in diameter and 4 in. high. The lower end of this pipe enters the heater 1 in. and the fumes of the lamp pass off through it. As the pipe reaches above the hover, the fumes do not hurt the chicks. The total height of the heater is 5 in. and the hover top rests on it.

A COMBINATION BROODER AND COLONY HOUSE
JESSE R. HARRIS, UNIONTOWN, PA.

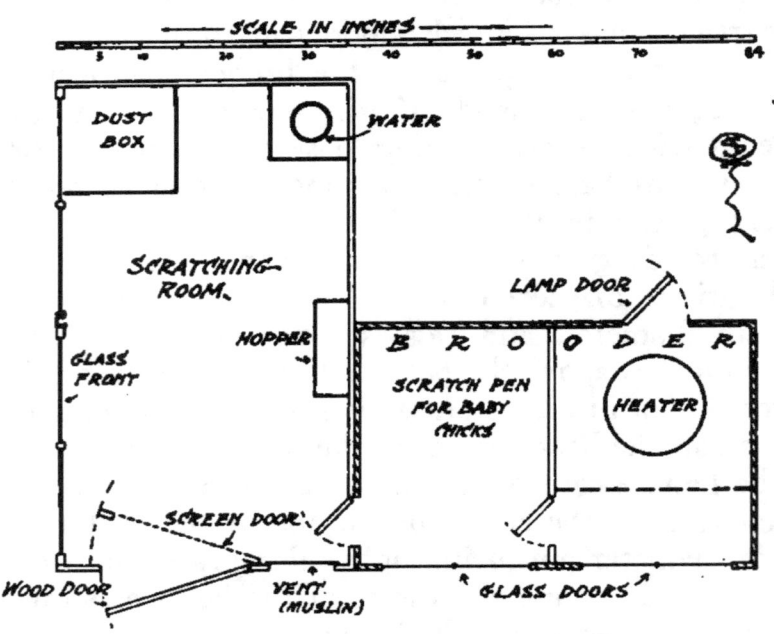

MR. HARRIS' COLONY HOUSE AND BROODER

85. *Feed and Attention of Chicks.* The floor is kept well covered with clean litter, and the chicks are fed by hopper after they are eight weeks old. The first year Mr. Harris only used one side of the brooder to hover the chicks; the other side he had for a scratch pen until the chicks were three weeks old. They were then admitted to the scratching room of the colony house. The second year he used both sides of the brooder to hover the chicks, and admitted them to the colony house when they were one week old. He let the two flocks run together and separated them at night. In the far right-hand corner of the house where the sun seldom strikes, is the drinking fountain. In the other corner of the same end where the sun shines most of the time, he has a dust box. He fills the box with ashes and mixes a little potash with it — this keeps the lice off the chicks.

CHAPTER V.

HOUSES FOR MALES OR SMALL PENS

86. *The Use of Male Houses.* As soon as a breeder of stand-ard-bred poultry gains a reputation for high quality stock, it is necessary for him to keep a supply of male birds on hand to fill the requirements of his customers. Females can be penned together, but male birds for sale or for exhibition purposes should be kept separate. Many of the successful breeders of the best strains of standard-breds mate a number of breeding pens of one male and one or two females. These pens are mated after careful study of the features and parentage of the birds, and it is from them the breeder expects his choicest offspring.

87. *A Convenient Male House.* One of these male houses is shown. It is 12 ft. wide, and any multiple of 3 ft. long. There is a passageway 3 ft. wide along the north side, and the remainder of the house is divided into narrow pens 3 ft. wide. Each pen houses one male, so that a house 60 ft. long would take care of twenty. The sills are made of 4-in. square stuff, if the house is not over 30 ft. long, and 4 x 6-in. lumber for a longer house. They should be painted with hot tar and placed on stones, or short posts. The lengths of the studs are: Center, 7 ft. 6 in.; rear, 6 ft.; front, 5 ft. The studs are set up 3 ft. apart.

88. *Labor Saving Devices.* The partitions between the pens are boarded solid for 2 ft. The remaining 4 ft. is covered with 2-in. mesh wire netting on every three partitions. Each fourth partition is covered with unbleached muslin (instead of the wire netting) to prevent drafts blowing through the house. When the muslin is up, it can be painted with boiled linseed oil to preserve it. In the wooden part of every other partition, a hole 9 in. wide by 6 in. high is cut. The lower edge of the hole is 6 in. above the sill, and the north edge is 6 in. from the center stud. A $\frac{1}{8}$-in. board 9 x 12 in. is nailed on the lower edge, and on this board a galvanized iron pan 8 in. in diameter and 4 in. high is placed for water. One pan waters every two pens. In order to minimize

the labor of feeding, the pans for the mash are in the lower part of the doors. The same size round pan is used as for water, and the construction of the shelf is similar. The pans are filled with mash from the passageway.

89. *Doors for Several Purposes.* The doors to the pens are 3 ft. wide — the entire width of each pen. This simplifies the cleaning, and it also has the additional advantage of closing the passageway when any door is fully opened. Thus, if it is necessary to change a male from one pen to another, by opening two doors he can be driven into the desired pen with ease. If you wish to run one male with half a dozen hens, each in a separate pen, you can easily do it with this style of a house. That would eliminate the use of trap nests. The upper portion of all the pen doors is covered with 2-in. mesh wire netting.

90. *Wire and Muslin Frames for Windows.* There is a 1-in. mesh wire covered frame 2 ft. wide by 3 ft. high in the front of each pen. The upper edge rests against the plate. There is also an inner frame of the same dimensions covered with unbleached muslin. The inside frame is hinged and can be raised to the roof and fastened up on pleasant days. The frames are made of $\frac{7}{8}$ x 2-in. stuff. The 2 x 3-in. roost is 2 ft. long, and supported by two pieces $\frac{7}{8}$ x 3 in. These are hinged to the wooden part of the partition 15 in. above the sill. The roost is 10 in. from the partition, and rests on a front leg when in position.

91. *Glass Windows Can Be Used.* The frame of the house should be covered with 1-in. lumber and heavy roofing paper. In a cold climate it would be advisable to substitute windows for the hinged muslin frames, when the males belong to a breed with large combs. Do not have small openings in the front wall, for the birds to enter or leave their pens. When they can exercise outdoors the wire frames should be removed, and the birds can pass in and out through the window openings. The small openings commonly used damage a male bird's tail and plumage. This house is a convenience where the poultryman will keep many males in condition. It is excellent for other purposes, such as sitting hens, fattening chickens, keeping young chicks, etc.

92. *Mr. Burhans' Male House.* Mr. Burhans has designed a similar house which we will next refer to. It is the result of a dozen different ideas worked out in the past, and he finds that for

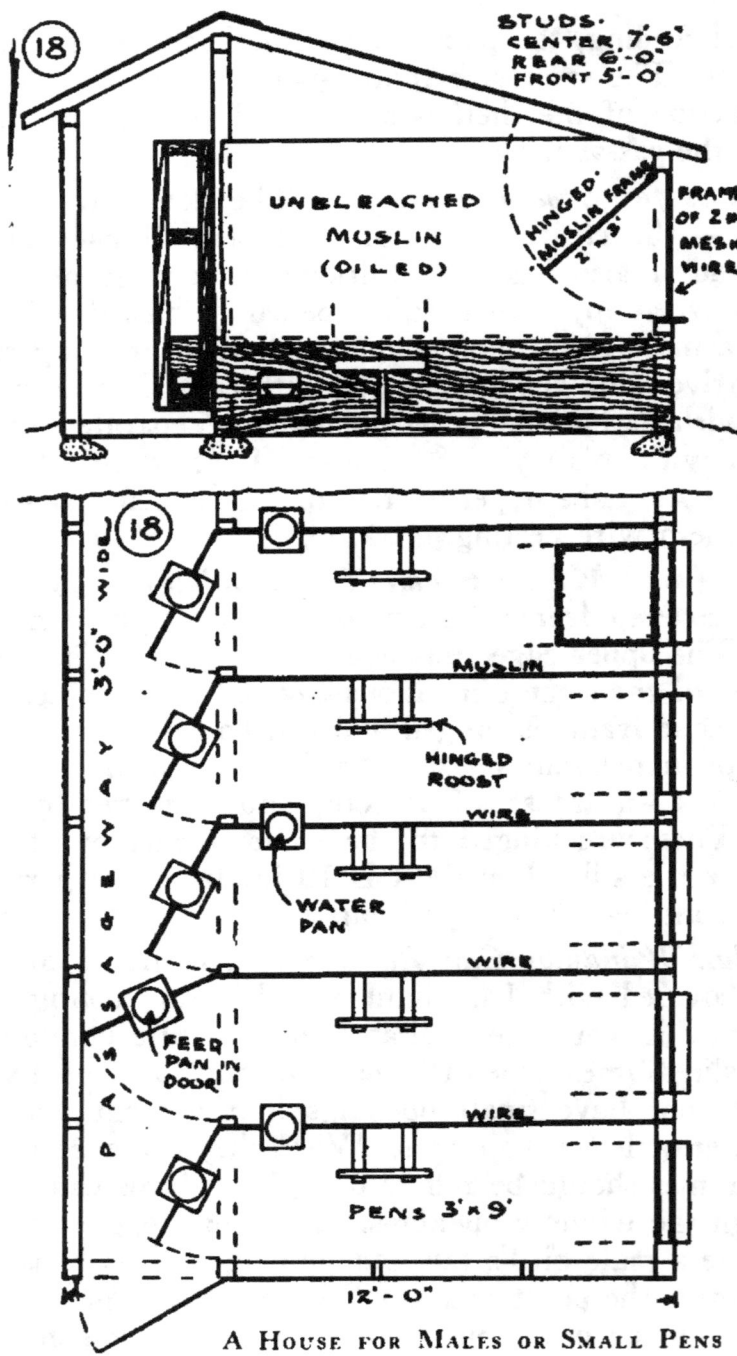

A HOUSE FOR MALES OR SMALL PENS

Upper — Cross-section of the male house showing one of the muslin covered partitions. Three-fourths of the partition are covered with wire netting above, instead of muslin.

Lower — Ground plan. Note that one water pan supplies two pens and that the feed pans are in the doors. All doors have spring (screen door) hinges.

being cheap, convenient, and adapted to other uses, it comes nearer being his ideal than any house he has ever seen. He uses it for matings during the early part of the spring, later for chicks and hens, then for summering over male birds after they are through with the season's breeding, and finally for fitting his show birds.

93. *Advice Thrown In.* "It is time and money gained to mate a couple of real good hens to a male, rather than to place eight more with him which do not happen to be suited to his characteristics." This statement is absolutely true, and the longer a breeder

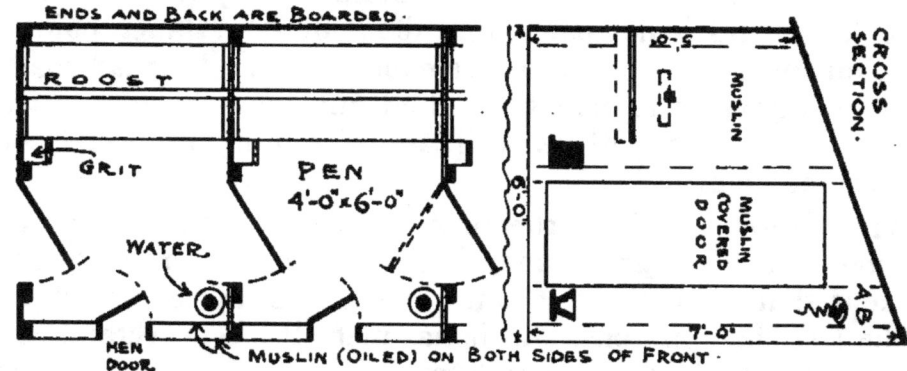

PLAN AND CROSS-SECTION OF MR. BURHAN'S HOUSE FOR MALES

follows the standard-bred poultry industry, the more firmly is it implanted in his mind. Another statement of this author is also borne out by the facts. "Give the chicks from two hens the room that the chicks from ten would take up, and the produce of the two good ones will cash in in the fall for more than all those raised from the pen." This, of course, applies only to high-quality standard-bred poultry, where the quality of type and plumage determines the value of the specimen. In chickens sold for table purposes, it would not apply.

94. *The Front is Made of Oiled Muslin.* The Burhans' house has a muslin front (oiled) tacked on inside and outside the studs. The ends and back are boarded. The floor space of each pen is 4 x 6 ft. The doors in the partitions are on double spring hinges. An exhibition coop for a single bird may be placed on the droppings board and two birds fitted in each pen. Going through the pens makes the birds very tame, and the size of the pen enables one to handle the birds easily, without having to chase them. There is a yard in front of each pen. The house is 6 ft. wide, 7 ft. high in front and 5 ft. at the back.

CHAPTER VI.

ADAPTING BARNS FOR POULTRY

95. *A Practical Question.* In the fifth issue of THE STANDARD the following question was asked THE STANDARD CLUB of poultrymen. "A beginner wants to obtain the greatest number of winter eggs, and is not particular about meat or fancy points. He has room for twenty birds in a remodeled stable that is bright, and can be made warm if necessary. Tell him what birds to buy, how to feed them so they will lay well, and give him any special pointers that will be helpful."

Mr. Jensen obtained the prize award for his answer to this question, and his plan of adapting his barn for the purpose is so excellent that we include it here. He has followed it with success for the last three seasons, and it has met all his expectations. Last year he cleared $26.96 from twenty-four birds that he wintered and sold on March 22nd to a farmer at 10 cents a pound.

96. *Mr. Jensen's Answer.* Starting with the supposition that your barn is at least 16 x 16 ft., we advise your dividing it into two compartments of 8 x 16 ft. each. Buy 2-in. mesh netting reaching clear to the ceiling. The west half arrange as follows. Make a crate 12 in. from the west and north walls; also along the dividing netting to reach about 8 ft. from the north wall. Then take a couple of boards and place them on top of the crate — this need not be more than 5 ft. high — and fill the space with straw up to the ceiling. This gives you a straw inclosure 6 x 7 ft. x 5 ft. high. In this inclosure put your roosts not to exceed 2 ft. from the ground; also your nests and drinking water, if there is room. The other half use as a scratching room. Put in 12 in. of straw, which should be renewed every two or three weeks, also a dusting bath. Use fine sand, in which you can place insect powder or sulphur every eight or ten days. Although the care and feeding of poultry is rather out of place in a book of this kind, we will refer to Mr. Jensen's system, because it explains how the adapted barn had best be used.

97. *His System.* He recommends the selection of twenty Plymouth Rock pullets hatched early in March. Those are selected that stand high on good, heavy legs, clear eye and bright comb, with long, straight back and glistening plumage. They are placed in their home not later than November 1st, and fed sparingly during the first half of that month on equal parts of wheat and oats (one-eighth gallon twice a day) and 10 to 16 ounces of green bone every second day. They are allowed to run outside as long as the weather is fit. Cold will not hurt them as long as it is dry. If they are inclined to stay inside, they are chased out and the barn shut. This is absolutely necessary in order to have the pullets laying during the winter months. If you have to chase them out, see to it that they can get out of the wind. A canvas screen 3 x 8 ft. put on the ground in L-shape is about as good as anything Mr. Jensen has found. He is very positive on keeping the pullets outside and adds: "Don't let them into the scratching part of the barn as long as they don't lay, and it is not too wet outside."

98. *Feeds Heavier in November.* Toward the end of the month start to increase their feed, adding corn on the cob to the same — from four to six fair-sized ears ought to be sufficient — given at noon each day. A hopper should be filled with grit and oyster shell, 2 parts grit and 1 part shell. On December 10th to 15th the first eggs should appear, each pullet laying four to six of the first cluster, when she will probably stop until December 20th to 30th. They should be laying ten to twelve eggs every second day if it is cold and stormy. If cold weather, twelve to fifteen each day from February 1st to April 1st. On the first day of April they should be sold to the butcher as they have about outlived their usefulness.

99. *Method of Feeding.* Mr. Jensen feeds these laying birds as follows: Morning 2 quarts of warm skim milk, 1 quart of shorts and 1 of bran, well mixed. At noon they receive from 10 to 16 ounces of green cut bone. At 4 o'clock $\frac{1}{2}$ gallon of whole corn in the litter. After the birds have gone to roost, $\frac{1}{2}$ gallon of wheat and oats is scattered in the scratching shed to start them to work the minute they can see. A cabbage head, a few red beets, or a mess of boiled potatoes once or twice a week in place of the bone gives variety and is appreciated. When the birds show evidence of too much forcing, the mid-day meal is cut out.

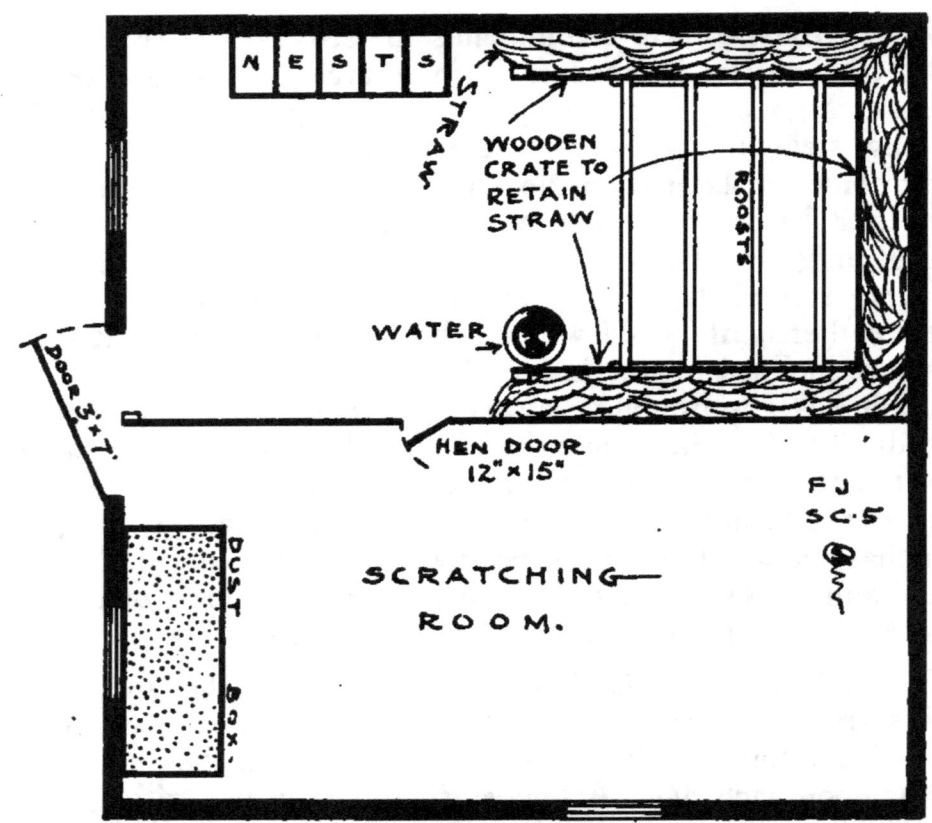

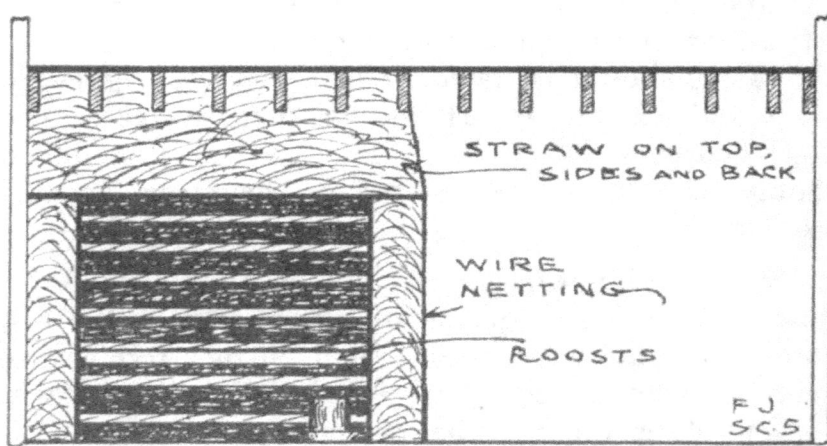

MR. JENSEN'S BARN FOR WINTERING PULLETS

The upper illustration is the ground plan, showing the roosting and scratching rooms divided by a wire partition. One large door gives entrance to both pens. The lower is a cross-section showing the straw on the sides, back and roof of the roosting quarters more clearly. Note the strips on the back wall to hold the straw in place.

100. *A Perfectly Dry House.* Mr. Rondeau has a ventilating system for poultry houses which he says cannot be beaten. He can leave straw on the floor for three months if he wished, and it would be as dry then as when he put it in. The house as illustrated is 30 ft. long and 12 ft. wide. It is divided into four pens; has a cement floor and a double-pitch roof. The ventilating is done as

MR. RONDEAU'S VENTILATING SYSTEM

This can be used in a barn or any style of poultry house and gives good results.

follows: Light wooden frames are made to fit in the four openings (*a*) where the chickens enter and leave the house. These openings are 12 in. square. The frames are covered with muslin and they are set in the openings during the winter. Overhead the boards are 2 in. apart, and 12 in. of straw is kept above them. Next, a hole is cut in each gable end. These holes are 2 ft. square covered with muslin (*b*). The house has three windows in the front, and one each in the east and west pens. The latter windows have muslin in the lower part of the sash (*c*) instead of glass. The door is in the west end near the back of the house. The droppings boards are along the north side. This is Mr. Rondeau's system complete and he kept twenty White Wyandotte and ten Single-Comb White Leghorn pullets in the house from October to December 31st last. These birds laid 1,823 eggs during that time.

CHAPTER VII.

MOVABLE FENCES AND FENCING

101. *Fences Built From Wire Netting.* The most popular poultry fence is made of 2-in. mesh wire netting 4 ft. wide, nailed above 2 ft. of boards at the ground. The latter is necessary, as we referred to in inside partitions, to prevent the male birds fighting. Do not, therefore, use wire on your poultry fence at the ground, if there will be a male bird on each side. This fence (6 ft. high) is sufficient for Minorcas and probably Leghorns, but for the American breeds 5 ft. high would be satisfactory, and for the Asiatics 4 ft. high. A woven wire picket fence 4 ft. high with building paper tacked along the lower half is excellent for the Asiatic breeds. Building paper can be tacked over wire. The top of the fence must not have a scantling or board of any kind running along it, otherwise the birds will fly up on the board and escape. A piece of wire $\frac{1}{8}$ to $\frac{3}{16}$ in. in diameter strung along the tops of the posts, with the top edge of the netting wired to it, will prevent the fence from sagging between the posts. The latter should be about 3 ft. in the ground and 16 to 20 ft. apart.

102. *A Better Fence.* A more durable and handsome fence can be made of the regular wire fencing that is about 1½-in. mesh at the bottom and gradually gets larger toward the top, the top mesh being about 4 in. wide. This fence is made of much heavier wire than the common netting, and it will hold its shape under all conditions. There is also a fencing made with diagonal meshes that some firms call "poultry and hog fencing." The mesh in this fence is about 2 x 4 in. and it is made of heavy wire that will last.

103. *Portable Fencing for Chicks.* Mr. Fuller has an original idea for confining small chicks that illustrates how larger sizes of poultry fencing could be used. He says that the best portable fence for growing chicks is 24 in. wide poultry wire fence. This can be bought at almost any hardware dealer at a lower cost than a fence can be built from other materials. The fencing is 1-in. mesh at

MR. FULLER'S DESIGN OF PORTABLE FENCING

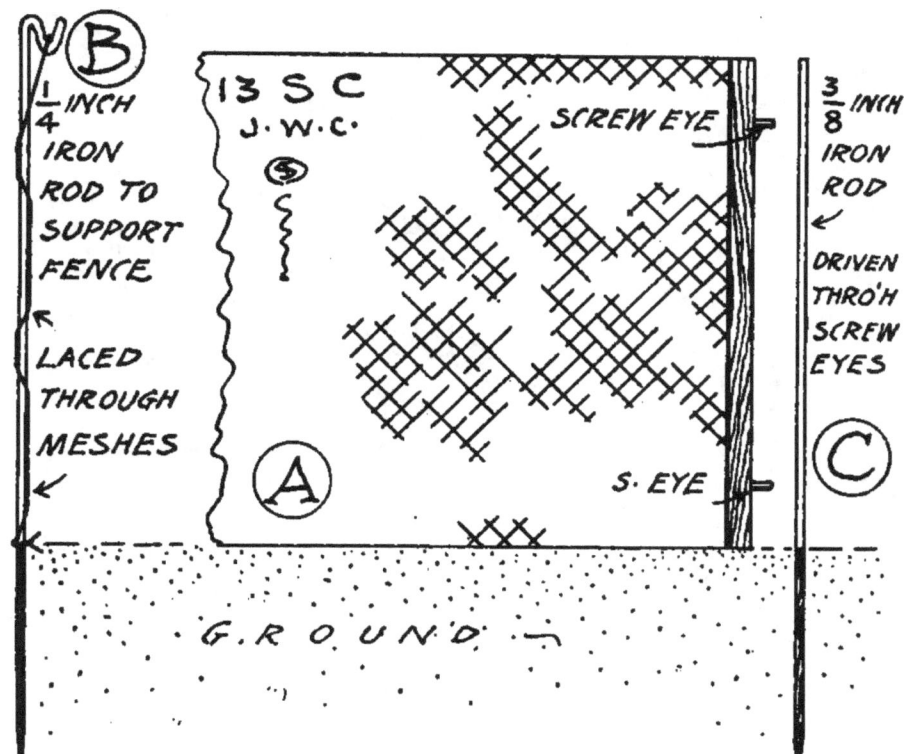

MR. CRUZAN'S FENCE MADE OF NETTING

the bottom and gradually gets larger until the top mesh is about 3 in. wide. Obtain it in any desired length, and attach it to 1 x 2-in. wooden cleats 3 ft. long. The fencing is fastened with small wire staples, and the cleats are sharpened at one end to drive into the ground. The cleats are nailed to each side of the brooder

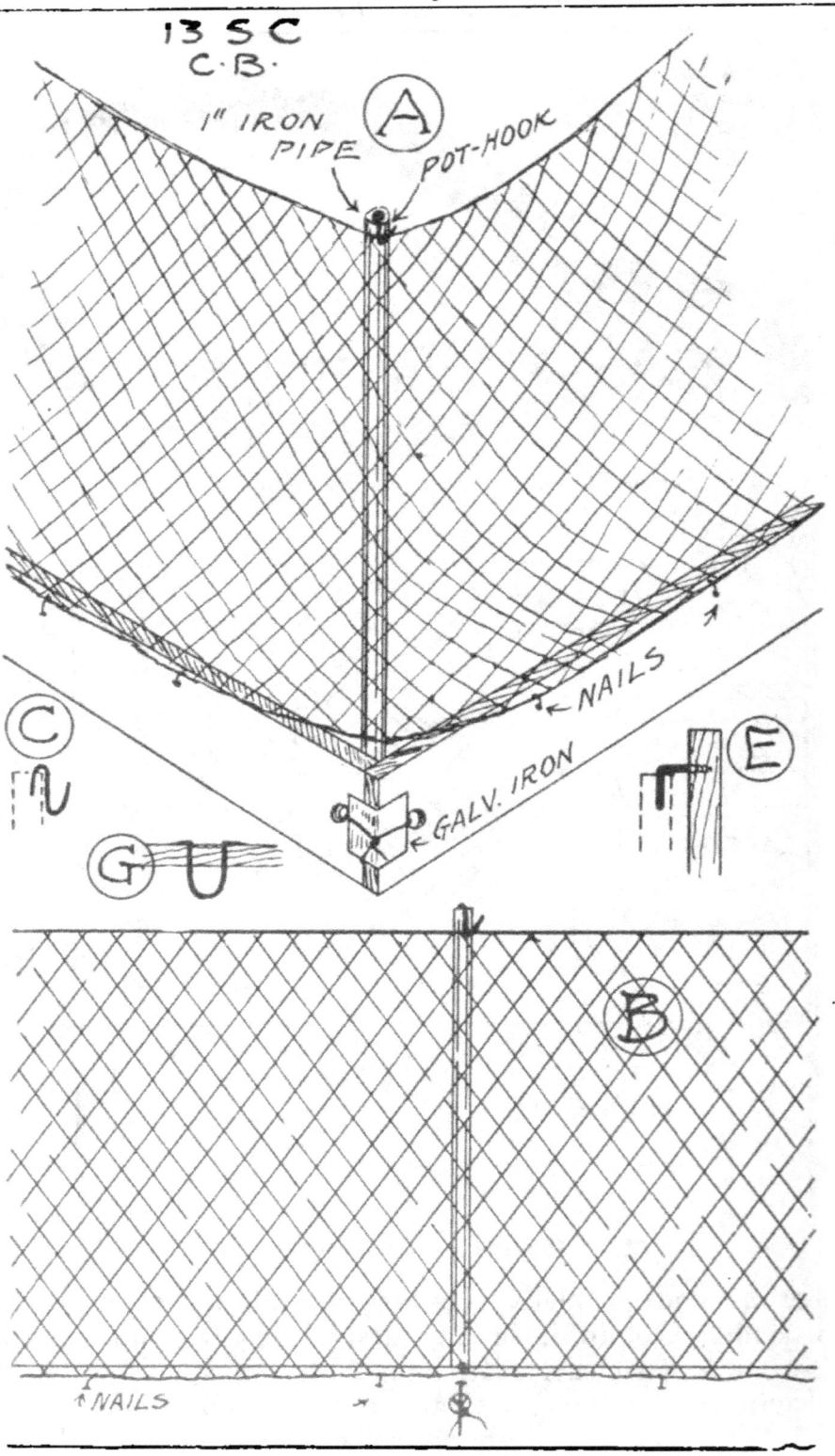

13 5 C
C·B·
1" IRON PIPE Ⓐ POT-HOOK
Ⓒ
Ⓖ U
← NAILS
GALV. IRON
Ⓔ
Ⓑ
↑ NAILS

or colony house, after being driven 12 in. in the ground. Other wooden stakes can be driven down where it is necessary to keep the fence in position. This fencing will make a square or round run of any shape or size. When not in use, pull up the stakes, roll

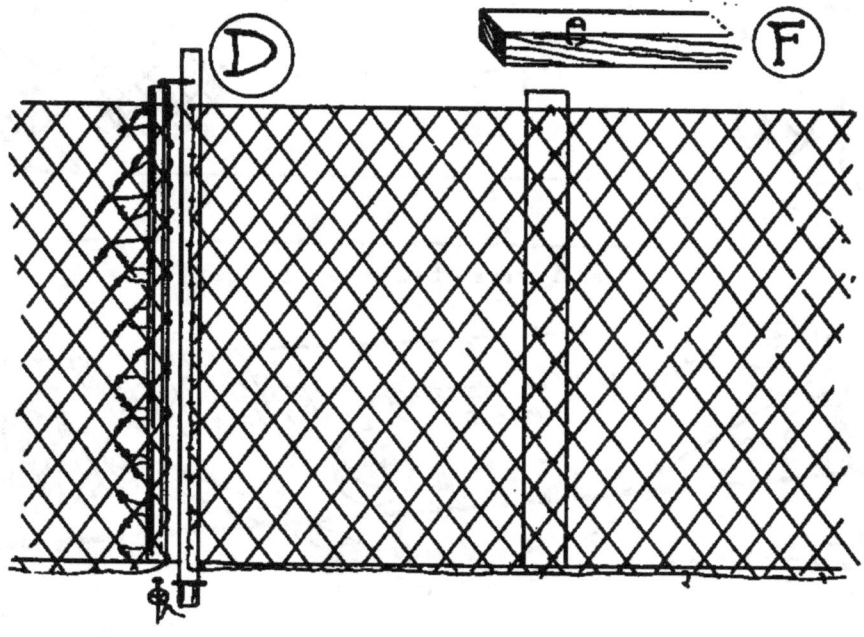

Mr. Batchelder's Fence with Iron Posts
Showing the convenient gate made without hinges.

up the fencing and store it away in a dry room. Mr. Fuller adds (and justly, too) that the straight wire fencing with perpendicular cross ties will stand erect, look well and is not expensive, whereas the common mesh poultry netting is too flimsy to stand upright.

104. *Netting Held Up By Rods.* Mr. Cruzan has devised a way to use the 1-in. mesh poultry netting 2 ft. wide, so that it can be kept in shape by iron rods. As in Mr. Fuller's plan, the netting is attached by staples to 1 x 2-in. end pieces 2 ft. long. Near the top and bottom of each end piece insert two screw eyes, of sufficient size to accommodate a $\frac{3}{8}$-in. iron rod. This iron rod is 3 ft. long and pointed at one end (C). For every 8 ft. of fence have a $\frac{1}{4}$-in. iron rod 3 ft. long, pointed at one end and with a hook at the other as shown in (B). This hook will hold up the fence. To set it up, take one of the 1 x 2-in. end pieces, stand it upright and put

[65]

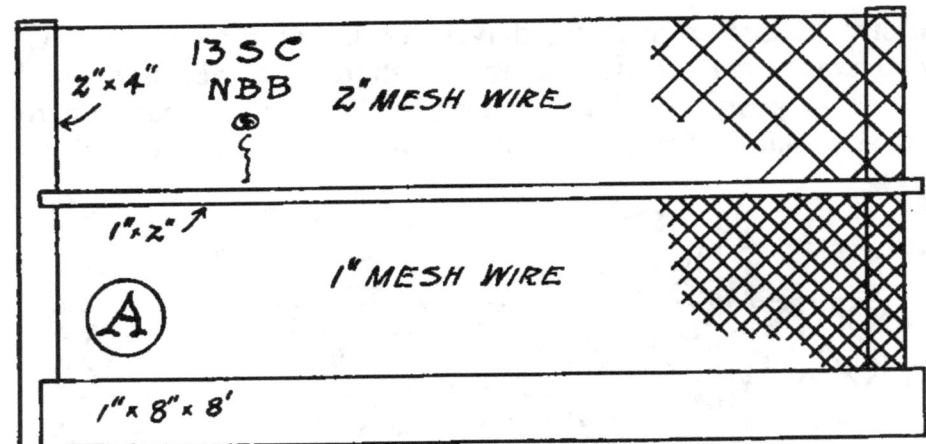

MR. BUTTLE'S SECTIONAL FENCE

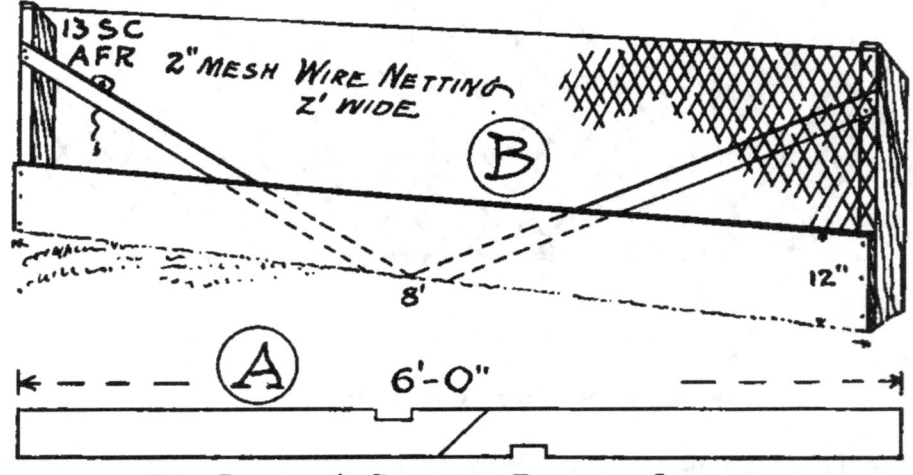

MR. RUEBNER'S PORTABLE FENCE IN SECTIONS

the $\frac{3}{8}$-in. iron rod through the screw eyes. Then drive the rod about 12 in. into the ground. Erect the other end similarly. Now take the $\frac{1}{4}$-in. rods and lace them through the mesh. Drive them into the ground; attach the top strand to the hooks and the fence is complete.

105. *Iron Pipe for Posts.* Mr. Batchelder uses a portable fence set up with iron pipe, except the wooden post where the gate is to hinge. One-inch pipe is cut into suitable lengths, after allowing for the height of the fence and one foot to enter the ground. At the corners, and where the gate opens, the pipe should enter the ground 18 in. If the ground is hard, use an iron bar to make the post holes, otherwise the posts may be driven. Use a hardwood block

with a hole bored part way through, to cap the iron posts while driving (F). Having set all the posts, make from heavy wire as many 1-in. pot hooks (C) as there are posts, and hang a hook on top of each. Have base-boards of the desired width and set them on edge around the yard outside the posts. Bore a hole in the boards opposite each post (B), except at the corner, where they should be bored 2 in. back from the ends (A). Pass a loop of soft iron wire around the posts and bring both ends through the hole in the board. Place a heavy nail across the hole outside (B), and draw the ends of the loop around this and twist the wire taut with pliers. At the corners a right-angled piece of galvanized iron is put over the ends of the boards as a protection for the wire (A).

106. *Erecting the Fence.* Take a roll of 2-in. poultry netting; wrap one end around the iron post where the gate is to open, and fasten it with its own wires (D). Carry the netting around the yard and hang the top strand on the pot hooks, keeping it as taut as possible. Arriving at the wooden post where the gate is to hinge, drive the staple to hold the netting in place. Then draw the wire down to the base-boards all around (except at the gateway) and fasten by small nails or staples, preferably the former as they are more readily removed. Fasten the netting with staples to the wooden posts in a straight line. Cut it where it meets the iron post on the left-hand side of the gate (D), and tack the end to a 1-in. square stick 12 in. longer than the netting is wide. Six inches of the stick should extend beyond the netting at the top and bottom. Drive a large staple (G) into the base-board where the gate will open. Place one end of the stick in this, and allow it to enter an inch or so. Drive an L-shaped hook (E) near the upper end of the stick to engage the top of post, and the yard is finished — gate and all.

107. *A Sectional Wire Fence.* Mr. Buttle has designed and used a sectional wire fence that is practical and inexpensive. The sections can be made almost any size, but a convenient size is 8 ft. long by 4 or 5 ft. high. To build the fence you require 2 x 4-in. uprights for posts; $\frac{7}{8}$ by 8-in. by 8-ft. lumber for the bottom rail, and $\frac{7}{8}$ by 2-in. by 8-ft. lumber for the center rail. Do not have a wooden rail at the top. The illustration gives the details, and you will notice that the sections are lap-jointed. A nail or a screw is all that is required where each rail laps, to fasten the fence se-

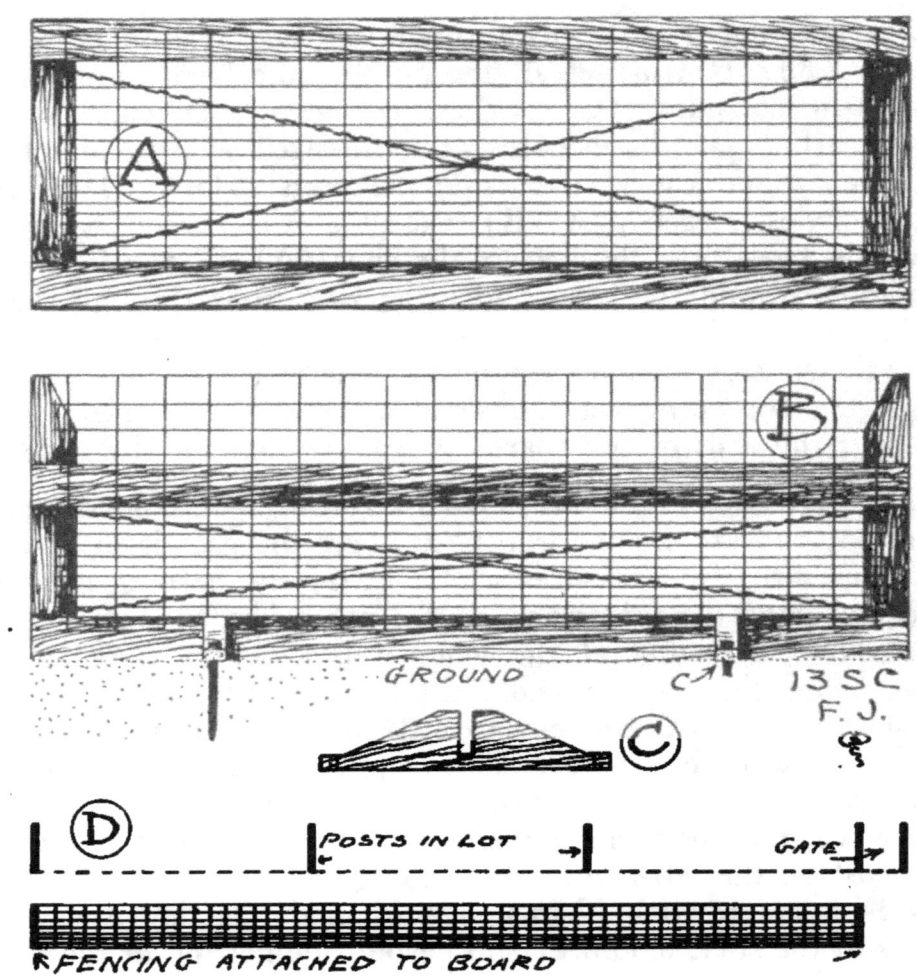

MR. JENSEN'S PORTABLE SECTIONS

The style shown in (A) was abandoned as the fowls flew over. Style (B) was satisfactory. (D) shows a unique idea for fencing a city-lot poultry plant.

curely. As shown in (D) the sections may be hooked together, if preferred.

108. *Mr. Jensen's Sectional Fence.* The following fence Mr. Jensen has found satisfactory for confining Barred Plymouth Rocks, even when they were full grown. Illustration (A) was his earliest attempt at building a movable fence, but he had to abandon it as the chickens always flew over. He then built the fence (B), and after he had pointed the end pieces, they never flew over again. The section is 16 ft. long by 4 ft. 6 in. high. The higher or brace board is 2 ft. 6 in. above the bottom edge. The two

brace wires are twisted in. Use a straight wire poultry fence, and stretch and fasten every wire — a claw hammer will draw it tight enough. You can drive stakes in the ground to hold it, but for greater convenience Mr. Jensen made pieces (C) out of 2-in. plank. Two of these are nailed to the bottom board of each section, 2 ft. from each end. If you fasten them closer to the ends, you cannot join the fence and turn a right angle. For greater security, drill holes at the ends of the piece, and drive two iron rods into the ground. This is hardly ever necessary unless the fence is placed in an exposed position.

109. *A Short Cut.* Mr. Jensen's yard is 50 ft. wide by 150 ft. long. During the breeding season he divides it into three different lots. The posts are set permanently, but the fence is used when needed and stored when the breeding season is over. He has a short-cut that fellow poultrymen in a similar position on a city lot will find useful. Forty-eight feet of poultry fencing is fastened to a board at each end. This strip of fencing is then laid across the lot (D) from a post at one side to the near post of the 2-foot gate. The two boards at the ends of the fencing are fastened by two nails to the posts. There are two other posts set 16 ft. apart between the division fence and the gate. The fencing is laid up against these and laths are nailed over it with a couple of small nails. Not a staple is used, and as the nails are not driven in far, the fencing is easily detached and rolled up for storing in the hay loft.

110. *A Simple Fence.* Mr. Ruebner has used a simple style of fence that is illustrated. A piece of 2 x 4-in. stuff 6 ft. long is cut in two slanting, as shown in (A). This makes two posts, where the netting is 2 ft. wide. Now you need a $\frac{7}{8}$-in. board 12 in. wide and 8 ft. long. Fasten the two posts to this board, and make two braces from 1 x 3-in. stuff, 4 ft. 9 in. long. They should be fitted and set down in the posts, with the other ends meeting at the center of the lower edge of the base-board (B). It is necessary to have the braces reach to the bottom of the base-board, as this makes a stronger fence and prevents it from warping.

CHAPTER VIII.

NESTS AND TRAP NESTS

111. *Location and Size of Nests.* The nests in poultry houses are usually placed under the droppings boards. This is done for two reasons: (1) to economize floor space; (2) to keep the nests dark and the curious birds away. The droppings board forms the top of the nests. Five or six are usually built together, and they are made without a floor and not fastened to the house. The inside dimensions are: 11 in. wide, 12 in. deep and 19 in. high. There should be a door 10 in. wide in the wooden front from which to collect the eggs. Common boxes of suitable size make good nests. The open side should be about a foot in front of the wall to keep the nest dark.

112. *Trap Nests.* Nests are called "trap" nests when there is some device by which the hen is confined to the nest after she has laid. By using trap nests, the poultryman is positive how many eggs each hen is laying, and also the eggs of each individual hen. We want to first refer to a nest used by Mr. Roberts, because it is the simplest trap nest we know, and it is not necessary to visit the house two or three times a day in cold weather to release the hens. This Roberts' nest does not distinguish the eggs of each hen — it simply separates the laying hens from the non-layers — and in the majority of cases that is sufficient.

113. *A Simple Trap Nest.* Mr. Roberts has used this home-made trap nest with good success. He builds a box 15 in. high, 12 in. wide and 3 ft. long. One end is left open with a teeter board extending 16 in. into the box. Just above the board a piece of poultry netting is stretched across the box. This netting is sufficiently wide to prevent the hen from coming back, but narrow enough to let her put her head under when entering the nest. Just in front of the nest proper there is an "exit" hole, out of which the hen walks into another yard or pen.

114. *Operation of Nest.* The hen walks in on the teeter board which lets down as she goes in, and rises into place as she steps off.

After she has laid she cannot return to the original pen, but must go out the side hole into another yard. At night all the hens that have laid are in one pen, and those that have not are in another. This method does away with the necessity of watching the birds that enter the nests. Above the nest, Mr. Roberts has a hinged door (not shown in the illustration) which allows him to gather the eggs and remove the nesting material with less trouble.

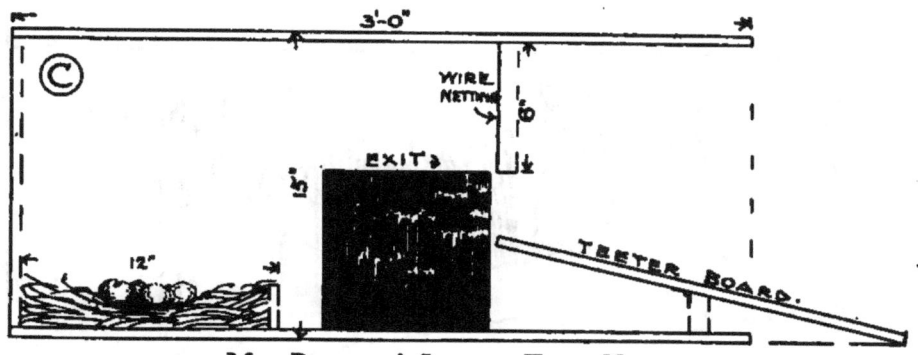

MR. ROBERTS' SIMPLE TRAP NEST

115. *Nest With a Wire Trap.* The following plan illustrates a trap nest that can be made from three pieces of wire, staples, hinges, and a box. This nest retains the hen until she is released. The box is 15 in. square and 2 ft. long. The top should have a door in the front half, by which the hen is removed to learn the number of her leg band. At the front end of the box, there should be an opening 10 x 12 in.

116. *The Trap.* For the trap you will need three pieces of heavy wire. One piece about 5 ft. long is bent as shown in (1). This is the wire door. It is attached by staples or screw eyes to the front of the box, inside and near the top, as is shown in the lower right-hand illustration. The shape of the wire door is governed by the size of the opening, provided you wish to install this trap in nest boxes already made. The two lower angles of the wire door should be sufficiently long to rest against the inside of the front, as shown. (2) shows the wire door resting in position, when it prevents the hen leaving the nest.

117. *The Trigger.* The wire door is held up, so the hen can enter the nest, by another piece of wire (3) called the trigger. It is hung out of plumb at the side of the nest, say 9 in. from the

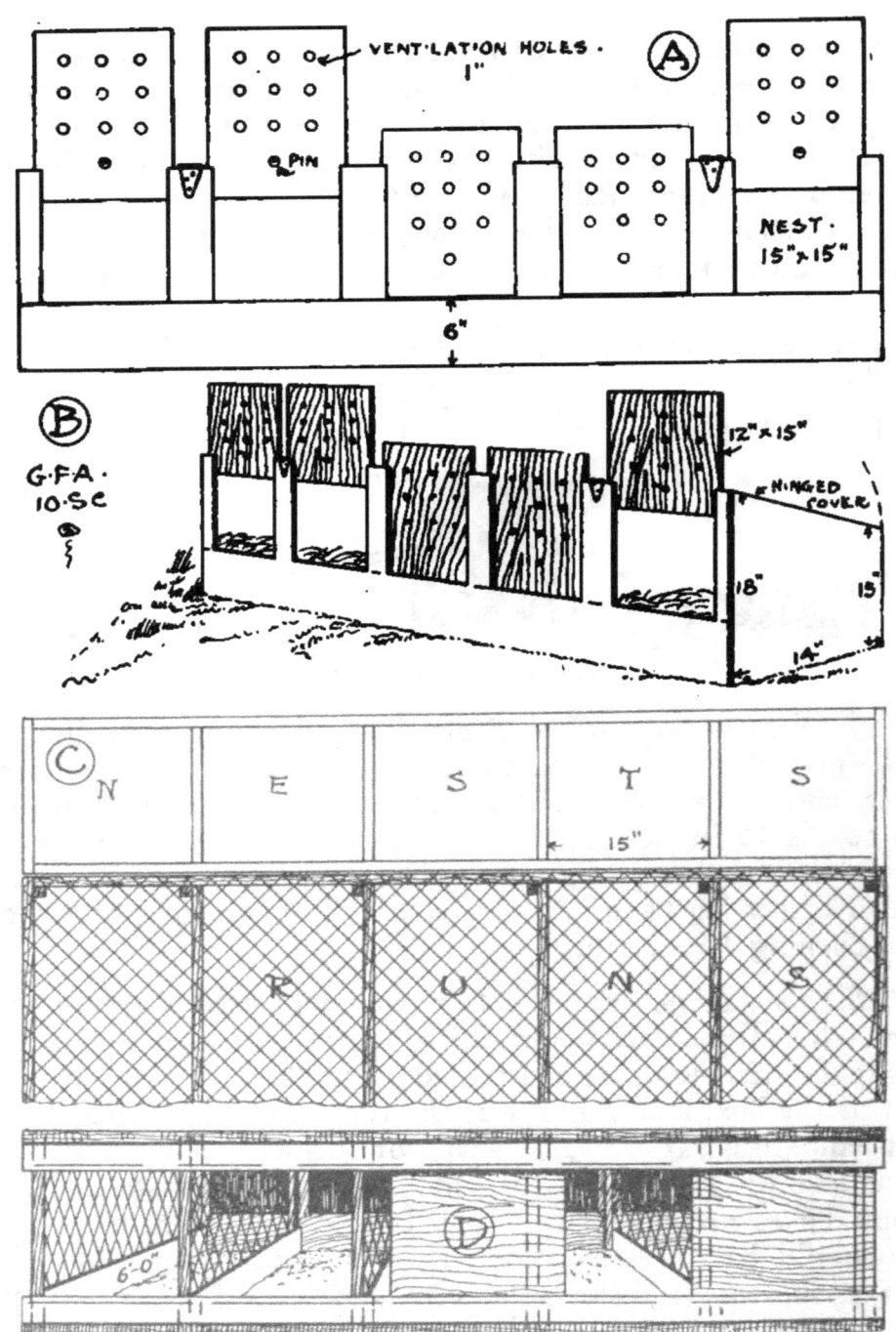

Mr. Albeck's Five-Compartment Nest for Sitters

(A) Front view of nest showing the sliding boards to confine the hens. (B) Drawing of the nest. (C) Top view of nest with the runs in position. (D) End view of the runs.

front at the top and 8 in. from the front at the bottom — in other words, it is hung so that it will fall and lay against the side of the nest when not in use. When the trigger is set it stands out at right angles to the side and extends three-fourths across the box. In the lower left-hand illustration, the wire door is seen supported by the upper wire of the trigger. When the hen enters she pushes the trigger aside and it falls back against the side of the box. This releases the wire door and it drops on the hen's back. As she continues forward it falls to a vertical position and closes the nest.

118. *The Lock.* However, it is necessary for us to have the other wire shown in (4) and called it lock, or it would be possible for several hens to enter the nest at the same time. The lock is wider than the opening and is attached to the front by four screw eyes, as you will see in the lower right-hand drawing. When the wire door is set, the lock rests on it, and when the door falls the lock must fall also. By referring to the two lower illustrations, you will note (a) the lock resting on the wire door, and (b) after it has descended and locked the door. Although it has required considerable space to explain this nest, it is easily built. Anyone can bend the three pieces of wire to the proper shapes, and have the nest working with not over an hour's work. A packing box of suitable size can be used, and the wires made to conform to it.

119. *Nests for Sitting Hens.* Nests that are used for sitting hens should be made without a wooden floor and placed around the sides of a pen reserved for this purpose. By all means keep the sitting hens by themselves, or they will be a continual source of trouble to you. The floor of the pen for the sitters should be earth properly leveled, and slightly moistened with water. A packing box 30 in. long, 16 in. high and 15 in. wide will make two nests. A wooden partition is placed in the center, and either a sliding board or hinged door is used in front, so that the sitters can be confined when necessary. Remove the entire top and the upper half of the boards of the back, and tack a piece of burlap on instead. This will give the hen sufficient air when the front is closed, and it will also keep the inside of the nest dark — a necessary point.

120. *Five-Compartment Nest.* Mr. Albeck in describing his five-compartment nests for sitting hens says that many failures in hatching chicks are due to setting the hens in the same house with

the other fowls. This results in broken eggs, a poor hatch and lousy hens. His plan of placing the sitting hens by themselves has given him great success and is thus described. The nests are built 18 in. high in front, 12 in. at the rear and 15 in. square. The runs

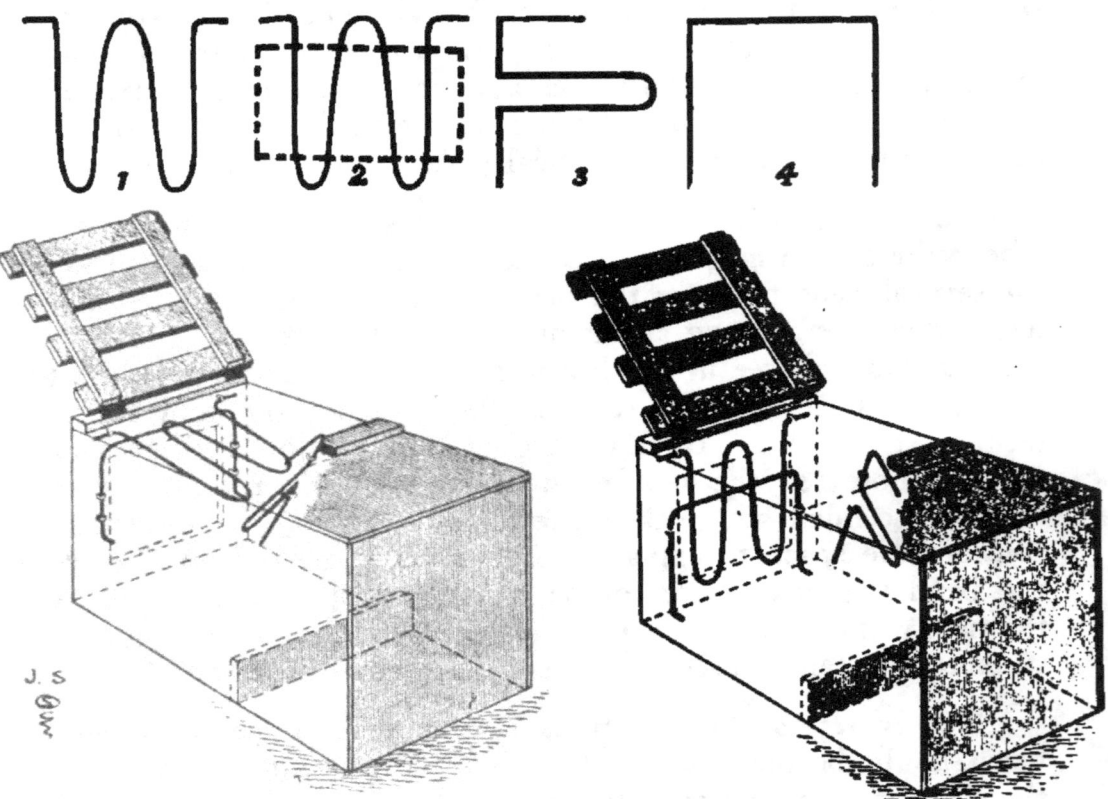

A SIMPLE AND RELIABLE TRAP NEST

The upper views are of the wire parts: (1) The trap. (2) The trap in position behind the opening. (3) The trigger. (4) The lock. The lower left-hand shows th nest with the trap set. The right-hand shows the trap down and locked, so that no hen can enter or leave the nest.

are 5 ft. long, or can be as long as you wish. A 6-in. board is placed at the bottom in front and a 4-in. board on the partitions, making openings for the hens of 11 x 12 in. A 12-in. board full of 1-in. holes for ventilation and made to slide up and down, makes the door to shut the hen in the nest. The top is hinged, so as to have an easy way to examine the hens on the nest, and to care for the broken eggs and chicks. We would recommend that the hinged top be made of a $\frac{7}{8}$ x 2-in. frame covered with burlap, instead of solid wood.

121. *Runs in Front of Nests.* The wire runs are frames 18 in. high by 5 ft. long covered with 1-in. mesh wire netting. They are held in place by stakes driven in the ground every 15 in., or just in front of each partition of the nest box, with another row of six stakes 5 ft. away (C). The ends are sliding pieces of boards 12 x 15 in. After the frames are set up, cover the runs with 1-in. mesh netting 6 ft. wide.

CHAPTER IX.

BROOD COOPS, RUNS AND PENS

122. *A Practical Coop.* Rearing chickens by natural means can be greatly simplified by having convenient, safe and easily cleaned brood coops. A practical coop should be so built that the chickens can be closed in at night and still have sufficient ventilation. It should be large enough to shelter the chickens during cold, wet weather. We include a brood coop described by Mr. Albeck which is convenient, safe and easily cleaned. The coop is built without a floor, but it rests on a wooden floor 3 ft. wide and 4 ft. long. The coop can, therefore, be lifted off the floor and the latter readily cleaned. We would prefer to have the floor fit inside the coop — and removable — so that the rain would not enter between the sides and the floor of the coop, as it would do in Mr. Albeck's coop. The roof is hinged at the front, and by raising this you can see how the chicks are doing and clean and limewash the inside without trouble.

123. *Construction of Mr. Albeck's Coop.* The front is built about half wood and half a wire-covered door. The wooden top and sides prevent rain beating into the coop — the sides protect the chicks in bad weather. The door is covered with ¼-in. wire cloth and it is impossible for mice, rates or other rodents to enter when the door is closed. The door can be held up in place by a heavy wire hook, or by a wooden leg hinged at the lower inside edge. When the top is raised the leg will drop into position and keep the door up. When the door is shut, the leg is folded up against it and rests between two of the slats.

124. *Coop in Use for Six Years.* Mr. Jensen has used the brood coop next described for the last six years and says he does not know how it could be improved. The frame is built of ripped 2 x 4's, and the coop is made of matched lumber from old boxes. The dimensions given are not always suitable for this kind of lumber, but they can be worked to as near as possible. The coop is 3 ft. long, 2 ft. 6 in. wide, 2 ft. 6 in. high at the front and 18 in.

high at the back. The floor is made of good matched lumber as is also the roof, it being covered with roofing paper. The roof is arranged so that it can be taken off for convenience in cleaning. Mr. Jensen has it with cleats on the inside and fastened down with two hooks, but you can hinge it, if you prefer. The door is 8 x 12 in. The slide clears a 4-in. opening 21 in. long. He also

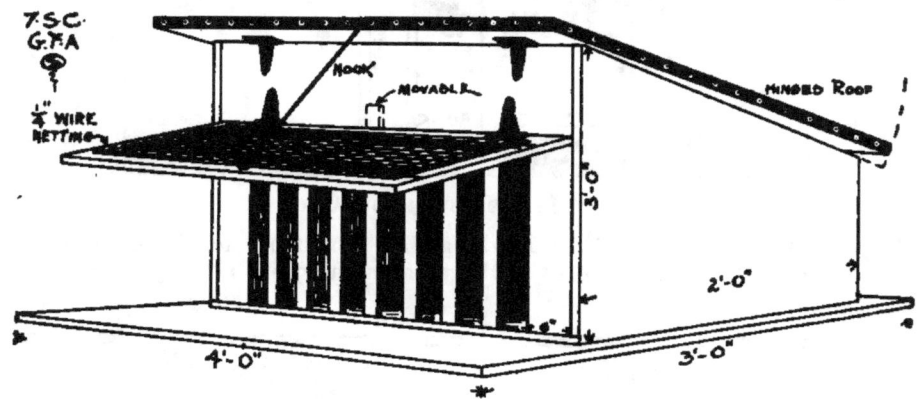

MR. ALBECK'S CONVENIENT AND SAFE BROOD COOP

The coop is not attached to the floor, but can be lifted and placed to one side. The wire-covered front door is a protection against rodents at night, while allowing ample ventilation.

uses wire screens like the screen doors and windows in dwellings. When the door and slide are closed, the coop is rodent-proof. The slide is used when you want to keep the hen in and let the chicks run; also when in the yard with the big chickens, you can shut the door and feed the little chicks without the larger ones being able to get at the food.

125. *A Lawn Screen.* Mr. Jensen also sends a plan for a lawn screen, or inclosed chick run. Two screens are made 8 ft. long by 2 ft. 4 in. high, and one screen 3 ft. long by 2 ft. 4 in. high. They are made of 1 x 4-in. stuff ripped for the end and top pieces. Halve them together and cover with 1-in. mesh wire netting. By putting hooks on the top and bottom you can hook them together, and with the coop for one side form the inclosed run.

126. *Coop Made of Roofing Paper.* Mr. Roberts has made quite an ingenious coop from netting and roofing paper. Two frames of 1 x 3-in. furring, 10 ft. long and 2 ft. wide are made; also two end frames 5 ft. long and 2 ft. wide. The end frames

could be 3 ft. wide for 'a smaller coop. These four frames are covered with 1-in. mesh wire netting. The top is also covered with the netting, with a door to lift up at one end. At the other end a shelter is made of a few extra pieces of furring and 2-ply roofing

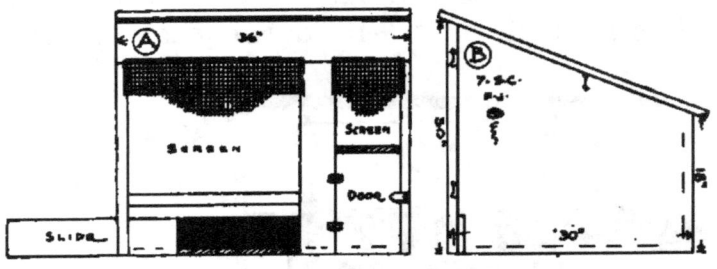

MR. JENSEN'S BROOD COOP

(A) Front elevation. (B) Side elevation. The front of the coop consists mainly of screens covered with common mosquito netting (wire).

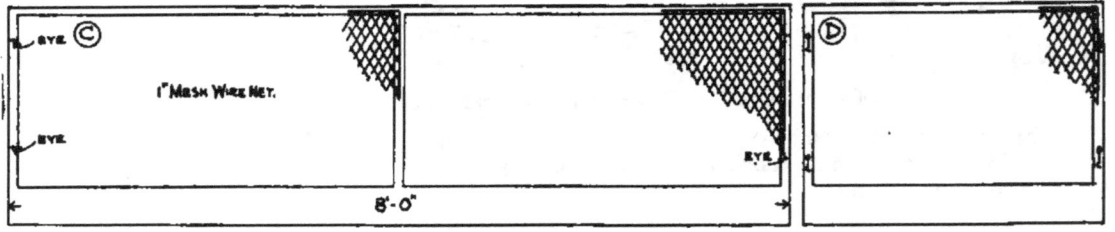

MR. JENSEN'S LAWN SCREEN

(C) Side elevation. (D) End elevation. The screen can be hooked to the front of Mr. Jensen's coop and makes a good sized run for the chicks.

paper. This coop will accommodate from 10 to 50 chickens and costs about $3.25 complete. Furring, 80 cents; roofing paper, 68 cents; netting, $1.60, and nails 20 cents.

127. *Roberts' Portable Fence.* Mr. Roberts also illustrates a portable fence made of the same strips of furring. The frames are 8 ft. long by 4 ft. wide covered with 2-in. mesh wire, and fastened together with hooks and eyes. The fence is easily moved and it will stand alone without any posts, if slightly offset, as shown in the illustration. Two sections can be built for $2.34.

128. *A Well Made Outfit.* Mr. Vandervort, on account of being very much troubled by his neighbor's cats and by hawks catching his young chickens, constructed several movable runs as illustrated. They consist of frames of 1 x 2-in. pieces securely nailed together. The sides and one end are covered with 1-in. mesh netting and the top with 2-in. mesh netting. A hinged door is placed in one end for feeding and watering the chicks. The runs are 12 ft. long, 3 ft. wide and 3 ft. high. This is sufficiently large for a flock of chicks until they are old enough to be out of danger of the cats, when they can be removed to colony houses. As the

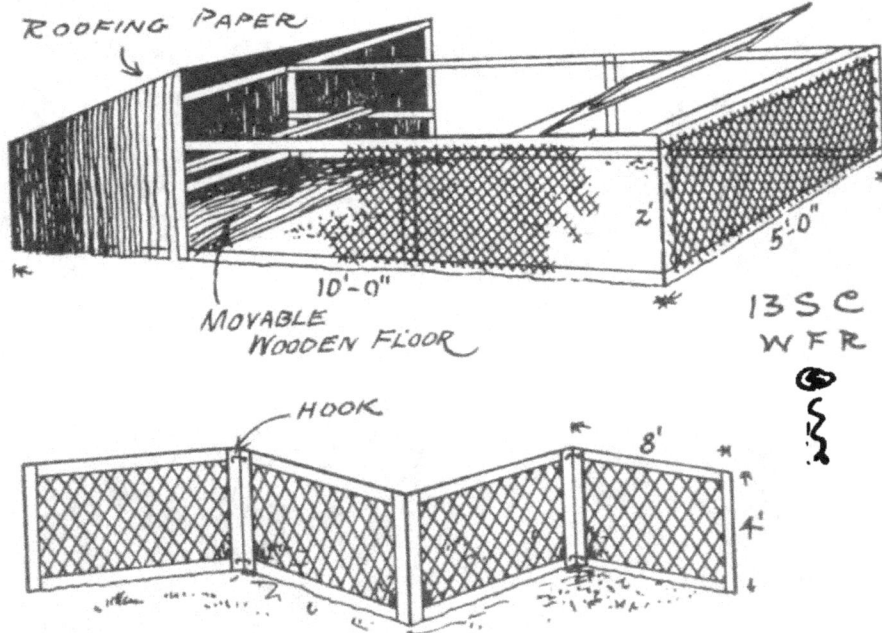

MR. ROBERTS' COOP AND FENCE

runs are light they are easily moved from place to place every few days. This gives the chicks fresh ground and insures good health. Mr. Vandervort believes that they grow as well as if they are allowed to roam about, and he is sure that they are always safe and sound from marauders.

129. *Mr. Crosby's Inexpensive Coops.* Mr. Crosby and Mr. Roberts have both the same idea of a practical, inexpensive coop. For a square coop, Mr. Crosby makes the sides 12 in. high and 6 ft. long — the end should be width of the coop. He also places a section on top of the run, and covers it entirely. The netting is fastened to slats 1½ in. wide and ¾ in. thick. This makes a satisfactory run (B). If you use A-shape coops, the runs are even easier to construct. Take wire netting the same width as the side of the coop and attach it to a frame. The two sections are then

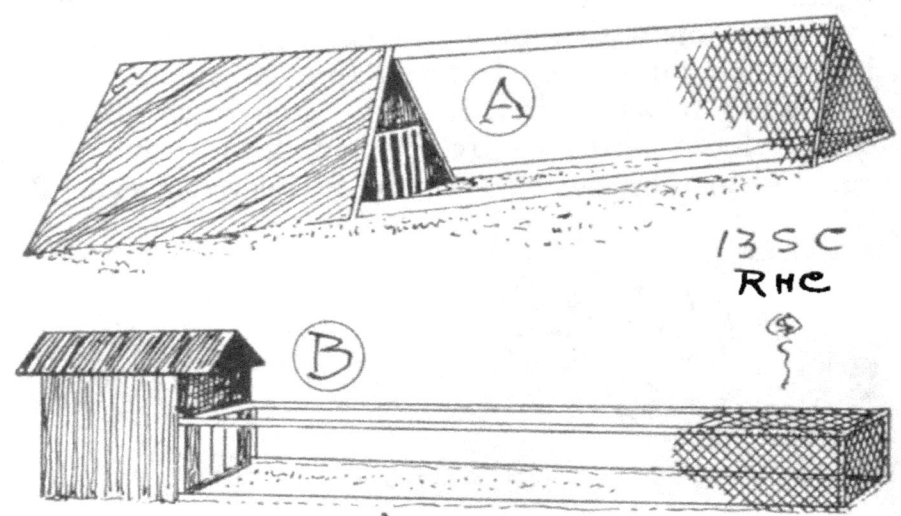

MR. CROSBY'S INEXPENSIVE COOPS AND RUNS

MR. THORNHILL'S COOP MADE FROM A SHOE BOX

set up on the same slant as the sides of the A coop. Build an end the same size as the front of the coop. A yard is formed in this manner, without having to use a special frame for the top (A).

130. *A Coop Made From a Shoe Box.* Mr. Thornhill has made an economical and suitable outdoor coop for a sitting hen, or a hen and brood of chickens in the following manner. The coop is

MR. THORNHILL'S COOP OPENED UP

built from a shoe box, of which the inside measurements are: Length 48 in., width 20 in., depth 14 in. The coop contains a little over 6½ sq. ft. of floor space and its depth is ample for the hen to move about in.

131. *Construction of Coop.* Examine the top and bottom boards of the box and take off the worst, so as to have the best lumber in the floor of the coop. Next turn the box bottom side up and nail a strip of wood — an old plaster lath will do — across it to keep it from falling apart. Now take out the front end board — drawing the nails with a nail puller — and nail strips on the

THE BURGESS FRESH-AIR HOUSE THAT CAN BE BUILT FOR $1 PER HEAD

The above house is built after a combination of ideas gathered by Prof. C. H. Burgess of the Poultry Department, Michigan Agricultural College. It is 20 feet square on the ground, with southern front open; windows on the east and west as illustrated. Houses 100 birds. As full blue print specifications accompany this book, on request, we omit detailed description of the house. Suffice to say that we consider it one of the most satisfactory houses that can be built.

inside, flush with the front ends of the box. This will keep the sides together and serve as front cleats to keep the front wire screen in position. See that the box is nailed together securely and especially at the coop end. Next take a piece of board 6 in. wide and 18 in. long, and saw it across diagonally from corner to corner.

A SHELTERED COOP FOR CHICKS OR BROODER

A SECOND STYLE OF SHELTERED BROOD COOP

Fasten these slanting pieces to the rear end of the box so as to slant the roof. Now make your roof of the lid boards, cutting them about 22 in. long and cleating them at each end with 2-in. strips, so as to fit the roof·end of the box. Hinge it to one side so as to facilitate cleaning, etc.

Make a frame upon which fasten 1-in. mesh wire netting for the front of the box — to slide up and down as required. Then make a frame covered with 1-in. wire netting for the top of the box and hinge it to the side. If a little care is exercised in making these fittings, you will have a coop that is roomy, dry, comfortable and safe for the hen and her chicks. Take a board 12 to 15 in. wide and hinge it to the front of the roof, letting it rest on the

PORTABLE RUN FOR CHICKS.
F.F.AUSTIN·WOODBRIDGE·NJ·

Mr. Austin's Run for Confining Chicks

Old carriage rims or saplings can be covered with 1-inch mesh wire netting and made into serviceable runs.

wire frame on stormy days, and turning it back over the roof in pleasant weather. Nail roofing paper over the roof and your coop is complete, unless you wish to paint it as I did.

132. *Cost of Coop in Cents.* Box 17½, hinges 15, wire netting 7, hooks 3, nails 2, rubber roofing 6½ — a total of 50 cents. The boxes may be bought from your own shoe dealer. The uses are: (1) You can place the coop outdoors in the garden and set your hen in it. She will be dry and comfortable and free from molestation. (2) When the chicks are hatched and after the hen has been well dusted with insect powder and the coop thoroughly cleaned and disinfected, the hen and chicks can be returned and they will have a shelter and small yard all to themselves. Here they can scratch around until they are strong enough to have the run of the garden. (3) The hen and chicks can be confined on threatening, stormy days, or on any morning when you wish to prevent them running in the wet, dewy grass. (4) When the chicks are weaned the coops may still be used on the colony plan for small bunches of growing chicks in the absence of more roomy quarters; assuming, of course, that the coops are kept thoroughly clean and disinfected. (5) They are admirable coops to use with fireless brooders.

133. *Mr. Cavanaugh's Coop for the Garden.* An excellent coop for confining the mother hen and allowing the chicks the run of the garden has been built and used by Mr. Cavanaugh, the General Agent of THE STANDARD. The dimensions of the coop are given in the illustration, and its original feature is a sheet-iron front that slides up and down. This sheet can be held at any desired height by inserting a pin or nail into one of the center holes. The nail rests on the roof of the coop. It is well to fully lower the front at night and partially in wet weather. Notice the ventilator in the front, for use when the iron is lowered to the ground. This makes a dark coop in which the chicks will rest quietly until the attendant opens the front in the morning. If the front were glass, the hen would probably tramp some of the chicks in trying to escape before the poultryman came. Instead of the solid door at one end, we would prefer a hinged frame covered with burlap. This would increase the ventilation, and the coop would then be perfect for its purpose.

134. *Coops for Early Spring.* When chickens are hatched in early spring it is desirable to have outdoor coops that will protect them from sudden changes in the weather. Such a coop is illustrated, and it will be seen that half of the front can be covered by an ordinary window. The right-hand compartment of the front should be covered with 1-in. mesh wire netting (not shown in the illustration on account of wishing to depict the interior more plainly). The window is pushed up flush with the left-hand edge, and makes this side of the coop warmer than the other. The sleeping chamber is exposed by pushing window to right.

135. *Another Cold-Weather Coop.* A second style of sheltered brood coop is also shown. In this a 4-glass cellar window forms the most of one end, while the front is lathed. It would be well to have a frame covered with 1-in. mesh wire netting, and the same size as the front, to confine the chicks at night.

136. *A Run From Carriage Rims or Saplings.* Mr. Austin has made a portable run for rearing chicks, and has found it very useful. Four old carriage rims are fastened together 4 ft. apart by three 1 x 2-in. strips 12 ft. long. Two strips are nailed at each end of the rims near the ground and the other at the top. Place your wire over the rims and cut it the right length, so as to have

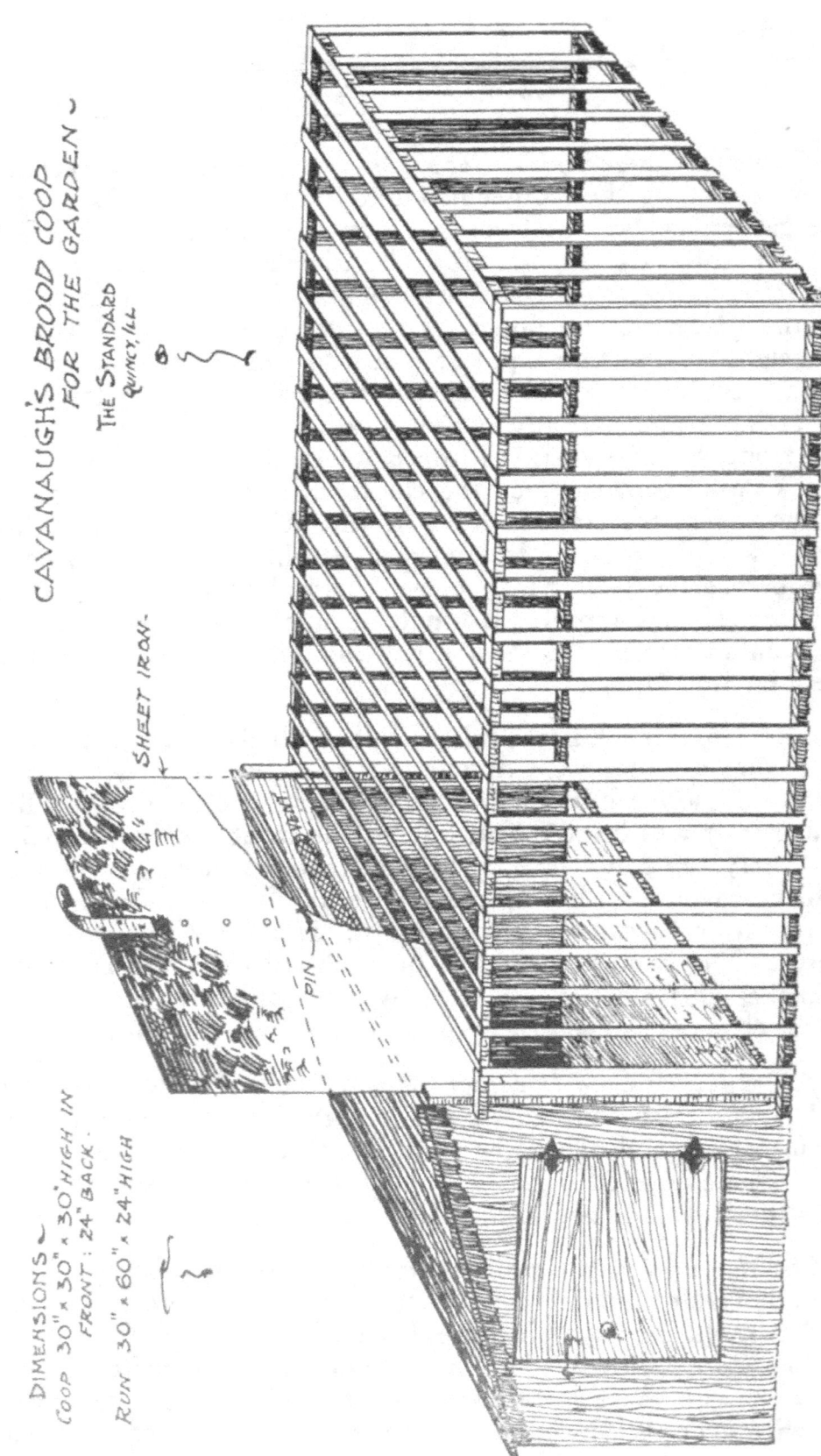

CAVANAUGH'S BROOD COOP FOR THE GARDEN ~

THE STANDARD
QUINCY, ILL.

SHEET IRON-

PIN VENT

DIMENSIONS ~
COOP 30" x 30" x 30" HIGH IN
FRONT: 24" BACK.

RUN 30" x 60" x 24" HIGH

MR. CAVANAUGH'S COOP FOR GARDEN USE

just enough to tack on the strips. One-inch mesh wire netting 6 ft. wide and 14 ft. long is used — the extra 2 ft. to close up one end. A coop of a hen and chicks is placed at the front or open end. A similar portable run can be made from four saplings 8 ft. long.

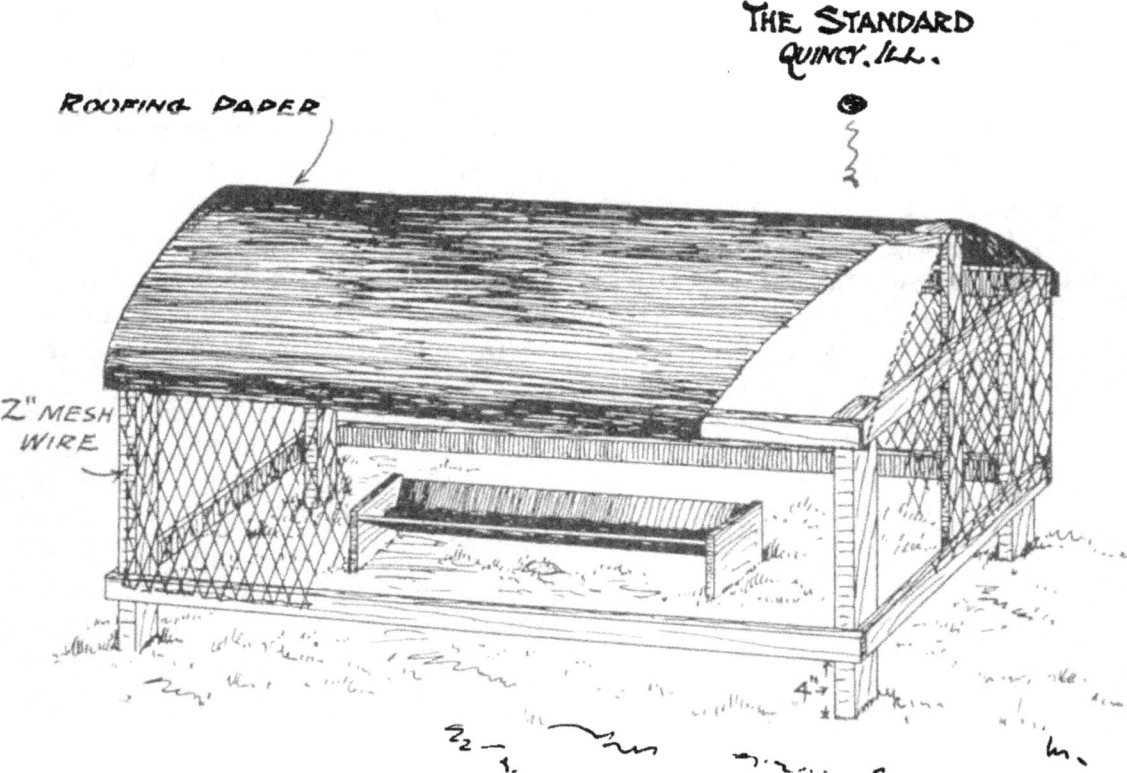

CAVANAUGH'S *FEEDING PEN FOR CHICKS*

THE STANDARD
QUINCY, ILL.

ROOFING PAPER

2" MESH WIRE

A PEN THAT PROTECTS THE CHICKS' FOOD

Get these in the woods; sharpen each end, stick one end 6 in. in the ground, and then bend the other end down and stick it into the ground. Keep the four ends on each side in a row and cover the saplings with wire netting in the same way as the carriage rims were covered. No strips are required when the saplings are used.

137. *Another Use for the Rims.* A feeding pen for small chicks can also be made from carriage rims. Use two rims and nail a few plastering laths on them, keeping the lowest lath 4 in. above the ground. Cover the upper portion of the pen with roofing

paper. In fastening the laths to the rims, do not drive the nails in the full way, so that the run or pen can be taken apart in the fall and the parts stored until the following spring. Mr. Cavanaugh illustrates and describes in the following paragraph the construction of a feeding pen of this kind that will show you how to use the saplings.

138. *Protecting the Chick's Food.* Mr. Cavanaugh's feeding pen was used for chicks that run with old fowls. The food trough is kept full at all times, and the chicks reach it by going under the bottom sill 4 in. above the ground. The top is covered with roofing paper and is, therefore, rain-proof. This is necessary, otherwise the food will be spoiled and wasted. The sides and end of the pen are covered with 2-in. mesh wire netting. This prevents the grown fowls eating the higher priced food intended for the chicks alone. It also allows the smaller chicks to obtain their share of food, without taking chances of being trampled by the larger ones. The only dimension in the pen that need be followed is the height (4 in.) of the bottom sill above the ground.

CHAPTER X.

MODERN FIRELESS BROODERS

139. *Advantages of These Brooders.* Fireless brooders are very rapidly increasing in popularity, and there is no doubt but that in a few years they will be the favorite type of brooder for raising chicks. It has been learned by experience that the chicks can be kept sufficiently warm in cold weather, and at other seasons of the year the fireless brooder is more easily attended to than a brooder heated by hot air or hot water. They cost nothing to operate; do not overheat or chill the chicks unless misused, and are get-at-able and easy to clean and disinfect.

140. *Mrs. Hinton's Desirable Brooder.* One of the simplest styles is used by Mrs. Hinton. She believes that fireless brooders are all right even in very cold weather, if the chicks have sufficient cover. For the older chicks a hover 36 in. square is made of three boards $\frac{7}{8}$ x 4 in. wide. The four legs are 2 x 4-in. stuff 8 in. long, nailed securely in each corner. On top of the boards three thicknesses of cloth are tacked. This cloth sags down in the middle so it will touch the chick's backs. Other pieces of cloth are tacked around the edge, so they will hang down to the floor. These are slashed in the customary way, to enable the chicks to go in and out of the brooder easily.

141. *"Growing Like Weeds."* Mrs. Hinton places her fireless brooders in colony houses, either 3 x 6 or 6 x 8 ft. From forty to seventy-five chicks are in each hover, and they grow like weeds. These brooders are well liked, because they are easier to clean than lamp brooders. A nest of straw is made under the center of the hover, and in cleaning, all that it is necessary to do is to raise up the hover, replace the straw with fresh, put down the hover, and the job is done.

142. *A New Wrinkle.* For very young chicks the hover is covered with a larger dry-goods box. In one side of the box saw out an opening 6 in. square. The box is then placed bottom side

1. MRS. H. P. HINTON'S

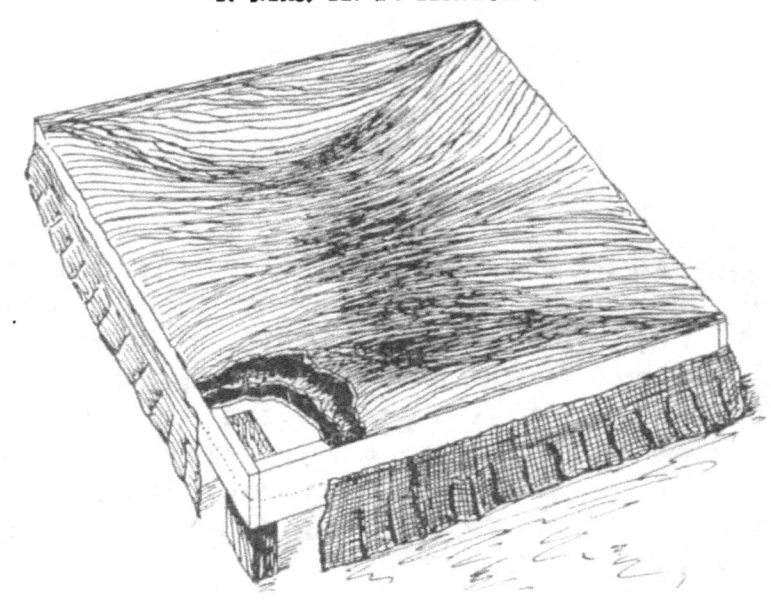

2. MR. J. W. CLARK'S

DIFFERENT STYLES OF FIRELESS BROODERS

up over the hover with the opening to the front, and there is then no danger of any of the little chicks getting cold — even with the temperature down around the 40's. In addition, when the weather is cold, Mrs. Hinton puts more cloths on the brooder, just the same, she says, as you put more covers on your bed when it is necessary.

"There is no excuse for one not having brooders in plenty, and these are something us women folks can make — I make mine. If one can't get boxes, they can be made from lumber, but it requires more work."

143. *Another Style.* Mr. Clark's brooder is 18 in. square by 12 in. deep. The cloth cover shown leaning against the wall in the illustration, fits into the brooder and can be fastened up by pegs at any desired height. There is a hinged wooden top to the brooder, which, when closed, protects it from the weather. Ventilating slides made of galvanized iron are on each side and in front, so that the brooder can be ventilated. It will be noticed the slides on the sides are at about the center of the box, while the front slide is near the top. There are three 6-in. square, hinged doors in front. For young chicks, one door is all that is necessary, and using only one keeps the interior warmer. The cloth top is made by tacking strips of old woolen material to a wooden frame. Several thicknesses of cloth are used, and the strips extend about 4 in. below the frame.

144. *Mr. Jensen's First Brooder.* Mr. Jensen first makes a frame 2 x 2 ft. 6 in. out of 1 x 2-in. stuff, then he takes plastering laths and nails on the frame an inch apart. Old sacks or woolen goods of any kind are torn into strips 1 in. wide and long enough so that they will hang down over the laths 3 or 4 in. The frame is placed into one of the compartments of a 3 x 6-ft. brood coop, so that it leaves a space of 6 in. around the hover on three sides. This makes it impossible for the chicks to crowd. Mr. Jensen advises the use of two hovers, as in this way you can clean and sun one while the other is in use. Raise up the frame as the chicks grow larger.

145. *A Fireless That Had Admirers.* Mr. Schultz used a fireless brooder for the first time this year and was so successful with it that he induced several of his friends to use them. He gets a box of suitable size — about 14 x 20 in. and 6 or 8 in. deep — at the

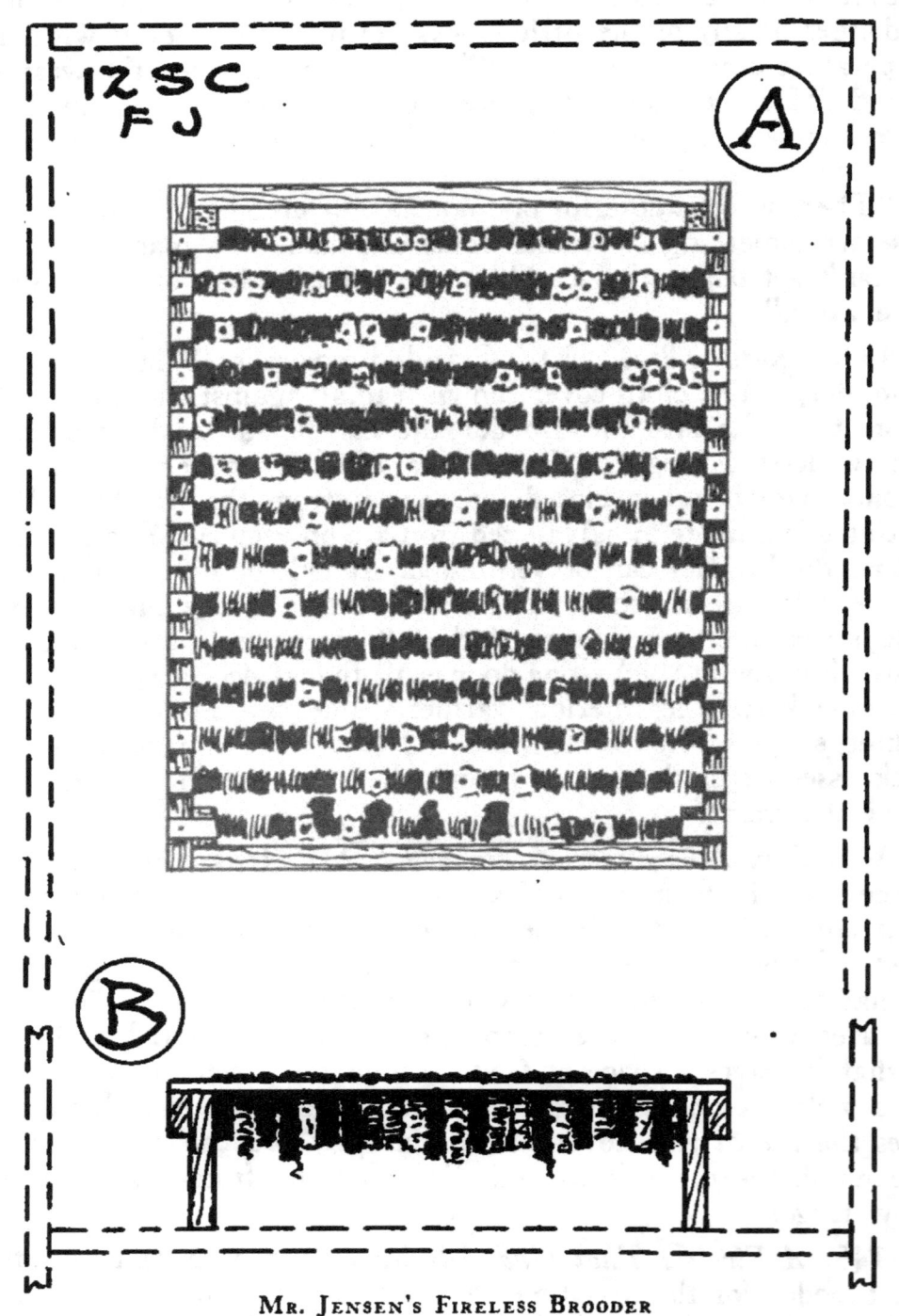

MR. JENSEN'S FIRELESS BROODER

store, and cuts a 4 x 4-in. hole in one side 1 in. from the bottom of the box. The piece cut out is made into a door, by using a scrap of leather for a hinge, so that the door opens down and forms a bridge for the chicks to get into the brooder. A wooden button at the top holds the door shut if desired.

146. *Further Details.* A frame of inch-square pieces is made to fit loosely into the box and a small nail is driven into each inside corner of the box, 5 in. from the bottom, on which the frame rests when in place. Next, a piece of heavy outing flannel is tacked on the frame so that it will bag down an inch or more. One-half yard of outing is what Mr. Schultz uses. Plenty of fine straw is put in the box to make a nest, so that the cloth will rest on the straw. With the cloth-covered frame in place, cloth side down, the brooder is ready for the chicks. This size is sufficiently large for twenty-five to fifty chicks. Do not crowd them.

147. *What the Fireless Did.* Mr. Schultz gives his experience with fireless brooders as follows: "I took twenty-eight chicks from the incubator March 27th and put them into a box made as described. I placed this box inside a large brooder box outdoors and raised every chick, although we had freezing weather nearly every night for several weeks. Since then I have used the fireless brooders exclusively, and now have my third lot in the brooder without any loss.

"These brooders have not cost to exceed 10 cents each besides my work, and I would not accept a lamp brooder as a gift, except to throw away the lamp and use the fireless hover. As the box has no cover except the cloth it must be put inside a larger box with a tight roof to protect it from the weather.

"I also put feed hopper and water inside the large box, so that the chicks need not go outside in bad weather. They require watching the first day until they learn to run into hover when cold, but it is surprising how soon they learn it and how comfortable they are when snuggled under their cloth mother. On cold nights when the chicks are young, I close the door of brooder to be sure they do not get out and also lay an extra thickness of cloth on the hover. Try this plan and you, too, will throw away your old lamp brooder."

148. *Another Opinion.* Mr. Roberts says: "The fireless, or lampless brooder has come to stay. I believe that more chicks have

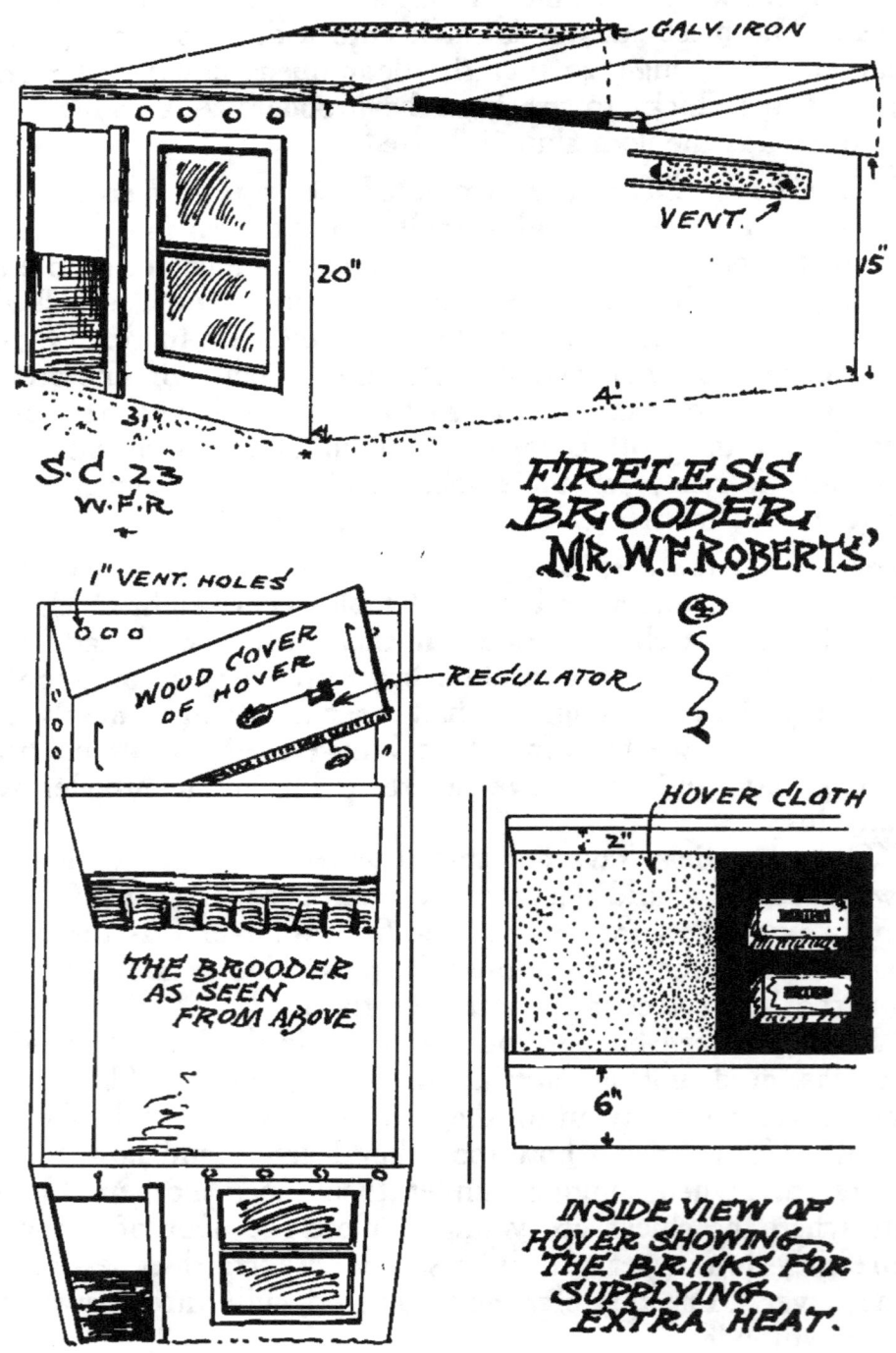

GALV. IRON

VENT.

20"

15"

31"

4'

S.C.23
W.F.R

FIRELESS
BROODER,
MR. W.F.ROBERTS'

1" VENT. HOLES

WOOD COVER
OF HOVER

REGULATOR

THE BROODER
AS SEEN
FROM ABOVE

HOVER CLOTH

2"

6"

INSIDE VIEW OF
HOVER SHOWING
THE BRICKS FOR
SUPPLYING
EXTRA HEAT.

Mr. Robert's Fireless Brooder

been lost by smoke and overheating in lamp-heating brooders than by being chilled. Accompanying is a sketch of a lampless brooder which I made and tested out to my satisfaction, and no more old hens or smoky lamps for me."

149. *A Fireless That Can Be Heated.* Mr. Roberts had a box 48 x 32½ in. He made the roof in two parts, hinged at the top of each with sufficient slant to shed the water. The brooder was then covered with roofing paper, and a window and door put in one end. Inside, he divided the box 21 in. from the rear end, the wooden partition extending from 5 in. from the floor to 1 in. from the top. From the lower edge of the partition a hover curtain reaches the floor. The hover frame is made of two boards 30 in. long and 6 in. wide. One board is nailed to the back 5 in. from the floor, the other at the same height 11½ in. from the back. Two pieces of 4½ x 1-in. stuff are nailed even with the bottom of the hover frames for hover supports. A hover curtain is tacked to the hover frame.

150. *Arranging for the Heater.* A box 4 in. wide is made to fit into this frame. To this box nail the hover blanket of flannel so that it will hang down on the chicks' backs. (In very cold weather an extra piece of flannel may be laid in on top of this.) Now get a piece of sheet iron 12 x 13 in., or wide enough to spring in tight to lay the heater on. This can be a soapstone, two bricks or a jug of hot water, either of which will keep hot for twelve hours. The wooden hover cover is made 11½ in. wide, 30 in. long; bore 2-in. holes for the regulator.

151. *Thermostat and Regulator.* Mr. Roberts uses a wafer thermostat and regulator in the cover. This controls the damper which lets off the surplus heat through the 2-in. hole. The regulator is set so that the damper will lift at 90 degrees. He also uses a piece of soapstone to increase the inside temperature of the brooder in very cold weather. The soapstone is heated on the kitchen stove morning and night. He gives his experience with the fireless in cold weather in the following words:

"I started mine the middle of last March with fifty-five chicks. I lost three the first week, two the second week. The rest of the fifty are all alive (except those that I sold for broilers) at this date, and a finer lot of chicks I never had. I have had three lots of

FIRELESS BROODER
AND COLONY HOUSE.
MRS. F.M.DOBBINS

BROODER.

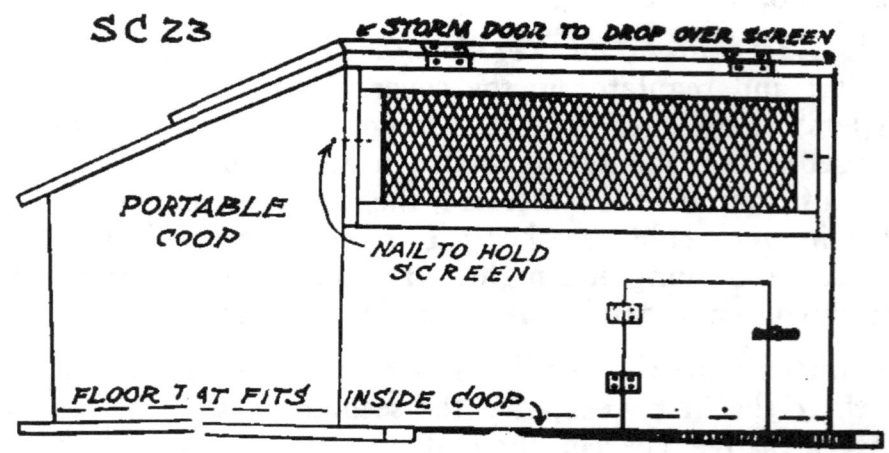

SC 23

STORM DOOR TO DROP OVER SCREEN

PORTABLE COOP

NAIL TO HOLD SCREEN

FLOOR THAT FITS INSIDE COOP

A PRACTICAL BROODER AND COLONY HOUSE

chicks in my brooder this spring — 181 in all — out of which I have lost twenty-two, which is good enough for me. Last year I hatched 154 chicks with hens and raised only seventy-three of them. As I said before, 'no more old hens for me.' "

152. **Another Simple Brooder.** Mrs. Dobbins' fireless brooders are 18 in. square and 12 in. high. They are made just like a box and then sawed through 6 in. from the bottom, and hinges fastened to the two parts at the back so that the top raises up like a lid of a box. There are six holes 1 in. in diameter in the sides at the top, three on each side, for ventilation. A door 4 in. square is cut in the front of the lower half of the box for the chicks to run in and out of the brooder. There is a frame made to fit inside the lower box just even to where it is sawed through. There is then flannel tacked over the frame so it drops 2 in. in the center. This is used for the hover. Then strips of wood are nailed on each side of the box for this frame to rest on.

Mrs. Dobbins tells how she operates the brooder and how she likes this system of brooding: "I keep hay chaff in the bottom part of the brooder, the amount depending on the coldness of the weather. If it is cold, I put in more than in warm weather. At night I put chaff over the door in front, leaving about an inch space at the top for ventilation. In very hot weather it isn't necessary to bank the doorway. You can tell by the way the chicks look when they come out in the morning, whether you are giving enough ventilation or not. If they come out damp, you may be sure you are covering the doorway too much. After the chicks are three weeks old the hover can be taken out.

"I think fireless brooders would be suitable for brooding chicks early in the spring, if they were placed in a building where one could have a little heat from a stove on cold, damp days and nights. I have used them outdoors, placed inside of a colony coop as early as March, with good success. I put twenty-five chicks in a brooder. The floor fits inside the colony coop, so water does not run in. It is loose from the coop so the box can be turned back, and the floor cleaned and sunned. I place a wire run in front of this coop 8 ft. long, 2 ft. wide and 1 ft. high. This keeps the chicks confined until four weeks old, then I turn them out to range."

153. *Changing Outdoor Brooders Into Fireless.* Mr. Vandervort explains how he converted a number of outdoor lamp brooders

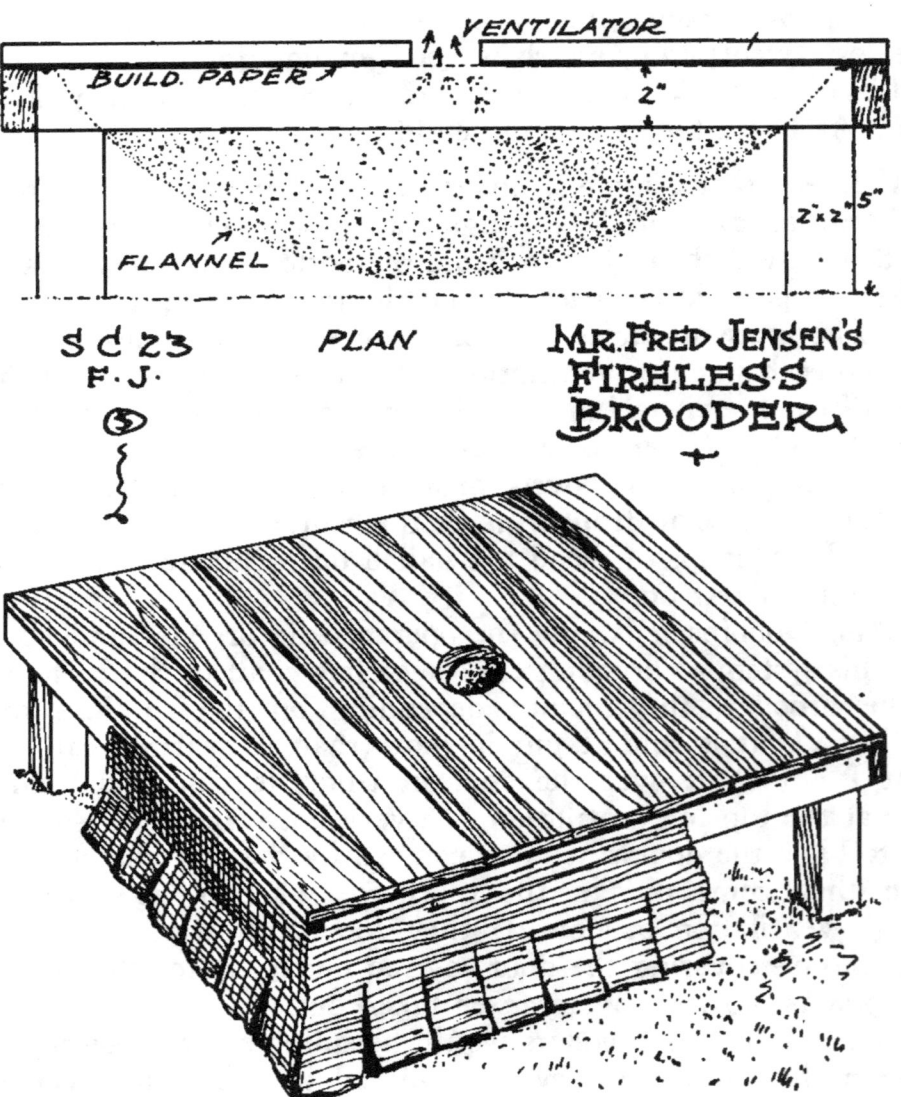

VENTILATOR

BUILD. PAPER

2"

FLANNEL

2"x2" 5"

S d 23
F. J.

PLAN

MR. FRED JENSEN'S
FIRELESS
BROODER

MR. JENSEN'S SIMPLER BROODER

into fireless. He did this in the following manner: "Four years ago I purchased six Cyphers style-A outdoor brooders. They had been used considerably, and that season became infested with lice, which I found were very difficult to get rid of. The next season I tore out the whole inside of them and applied a good coating of hot white wash and got rid of the lice. Three of them I used for brood coops and the others were made higher and used for colony coops.

[98]

"Having heard much about fireless brooders, I thought I would give them a trial last season, so I took one of the old brooders which I had used for brood coops, and converted it over into a fireless brooder. In one half of it and about 7 in. from the floor, I built a platform of matched ½-in. lumber. In the center of this I fastened the wire cylindrical chick-guard from one of the old brooders. Then I took one of the old circular hovers and tacked rows of slitted felt over its entire surface and about 1 in. apart. This was placed in position on the cylindrical chick-guard as in the old brooder.

"The holes in the top of the hover gave a good supply of ventilation. The other half of the brooder makes good space for feeding and exercising. From this to the top of the platform is connected a runway, which is fixed on hinges to close at night. This brooder proved very satisfactory; in fact, so satisfactory I am converting the other old brooders into fireless brooders.

"The first brood I tried in it was a brood of thirty Brown Leghorn chicks. They were hatched with hens, and after two or three days old were placed in the fireless brooder. I never saw chicks grow better and I only lost one of the thirty. In cold, damp weather the hover is taken out and thoroughly warmed, and also given a good sunning every day."

154. *Mr. Jensen's Second Fireless.* Mr. Jensen has also made a simpler style of fireless brooder than the one we previously described. In his latest model it is not necessary to use narrow strips and plastering laths, as can be seen in the plan of the illustration. Four pieces 1 x 2 x 24 in. long are nailed together to form a frame, on top of which is tacked a piece of building paper. Over the building paper are nailed thin matched boards. Under this frame are four legs so as to raise it 5 in. above the floor. Heavy outing flannel is then tacked underneath the frame so it would lack ½ in. of touching the floor in the center. Strips of flannel are tacked all around the edges of the frame, and afterwards slit every 3 in.

155. *Mr. Jensen's Opinion.* "Fireless brooders or hovers have come to stay at our place. We have several of them and they all work satisfactorily. During the month of March we had fifty-four chicks in a colony house 3 x 6 ft. and all the heat and hovering

they got after they were four weeks old was from a fireless brooder. The main thing to observe with this kind of hover is to set it away from the wall, especially a corner, so as not to give the chicks a chance to pile. When the chicks are eight weeks old or more, I use the lath hover described earlier. In that hover every bird can have his back covered and still have his head out in the open." (If the lath hover is to be principally used for the larger chicks, the laths should be about 2 in. apart instead of 1 in.)

CHAPTER XI.

HOT-AIR BROODERS

156. *Artificial Rearing is Preferable.* It is so simple to attend to several brooders, and especially when they are placed in portable houses, that even if we hatched our chicks under hens, we would much prefer to raise them by artificial means than by natural. It is a great convenience to have thirty-five chicks in one flock and to fully satisfy their wants in a warm, cosy brooder. While the sitting hen has her whims like any other animal, and sometimes refuses to hover her chicks when they are cold and tired, the hover of the brooder is always ready for them to run in and warm up. It is interesting to know how soon little chicks understand where the hover is, and what it is for. Even when they are a few days old, their instinct seems to tell them to run back where it is warm when they wish additional heat.

157. *A Good Hot-Air Brooder.* Mr. Deering uses an inexpensive brooder that gives him satisfaction. In our drawing we represented the upper section with a muslin top, although Mr. Deering has a wooden top to his brooder. The muslin is preferable in that it gives light to the interior and allows the foul air to pass out of the hover chamber. A special brooder lamp can be bought from companies that deal in poultry supplies, or an old-fashioned hinged burner will do. Buy the best burner you can with a 1-in. wick, and have a tin chimney 5 in. high made to fit it. A 1½-in. hole should be punched in the chimney and covered with mica in order to see the flame. The top of the chimney is placed directly under the lard pail and not less than 1 in. from the sheet iron. If the top of the chimney is close to the iron, the lamp will smoke and there is danger of its setting fire to the brooder.

158. *Construction of the Brooder.* A good stout box of a suitable size is purchased. The height of the lamp to be used (including the chimney) is measured, and the box is sawed in two, by sawing through the four sides. A section 2 in. higher than the lamp is cut off — measuring from the top of the box. This section

AN INDOOR BROODER
DESIGNED BY S. H. DEERING, MISSOULA, MONT.

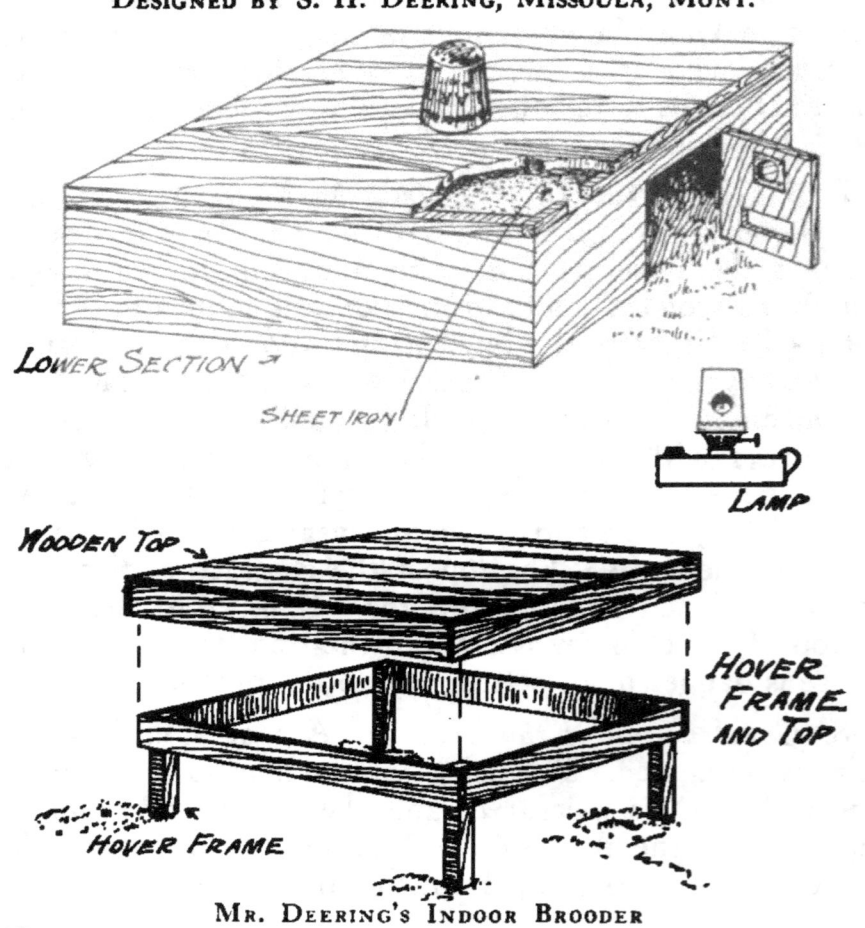

LOWER SECTION →

SHEET IRON

LAMP

WOODEN TOP →

HOVER FRAME AND TOP

HOVER FRAME

MR. DEERING'S INDOOR BROODER

when removed will be open at the top and bottom. The exact length and width of the section is measured, and a piece of sheet iron is bought that will just cover the top. Nail the sheet iron around the four edges of the section. On top of the sheet iron nail a strip of wood 1 in. square — through the iron into the sides of the box.

159. *The Heating Arrangement.* A wooden floor is then nailed to these strips forming a hot-air chamber 1 in. high under the floor. After you have the floor boards fitted, and before nailing them to the box, saw a round hole in the center floor board into which the open end of a 5-pound lard pail will fit tightly. This conducts the warm air from the hot-air chamber to underneath the hover. Before fastening the pail in place, punch a number of small holes in the bottom of the pail to allow the warm air to escape. Now fasten the wooden floor securely to the box. There should be an opening of 2 in. on one side of the hot-air chamber, in the 1-in. square strip, to allow fresh air to enter the heating chamber.

The sheet iron should be in one piece, or two if seamed together, so that the lamp fumes cannot enter the hover. Stove-pipe iron is heavy enough. In one side of the section cut out a lamp door. Near the bottom of this door have an opening to admit fresh air to the lamp, and near the top bore a 1½-in. "peep" hole. The hover sits on the floor with legs long enough to raise it 1 in. above the pail. The hover is like a small table but is made in two parts. A strip 2 in. wide is nailed around the legs and forms a table without a top. Now make a solid wooden top with the sides to come down 2 in. to cover the strips on the legs. The top must fit sufficiently loose to allow the hover cloth to go between.

160. *Copying a Fireless Feature.* Mr. Deering makes his hover cloth like the poultrymen who use fireless brooders. It is large enough to spread over the frame and hang down close to the floor; it must be so loose that it will sag down in the middle and rest on the chicks' backs. To allow it to do this, Mr. Deering cuts a round hole in the center of the cloth the size of the pail. A wire ring is sewn in this hole to fit around the pail. This lets warm air above the cloth and keeps it warm and comfortable over the chicks' backs — just like the old hen. This brooder is really a cross between a first-class lamp brooder and a fireless. It would be an excellent heated brooder for cold weather, and as the weather

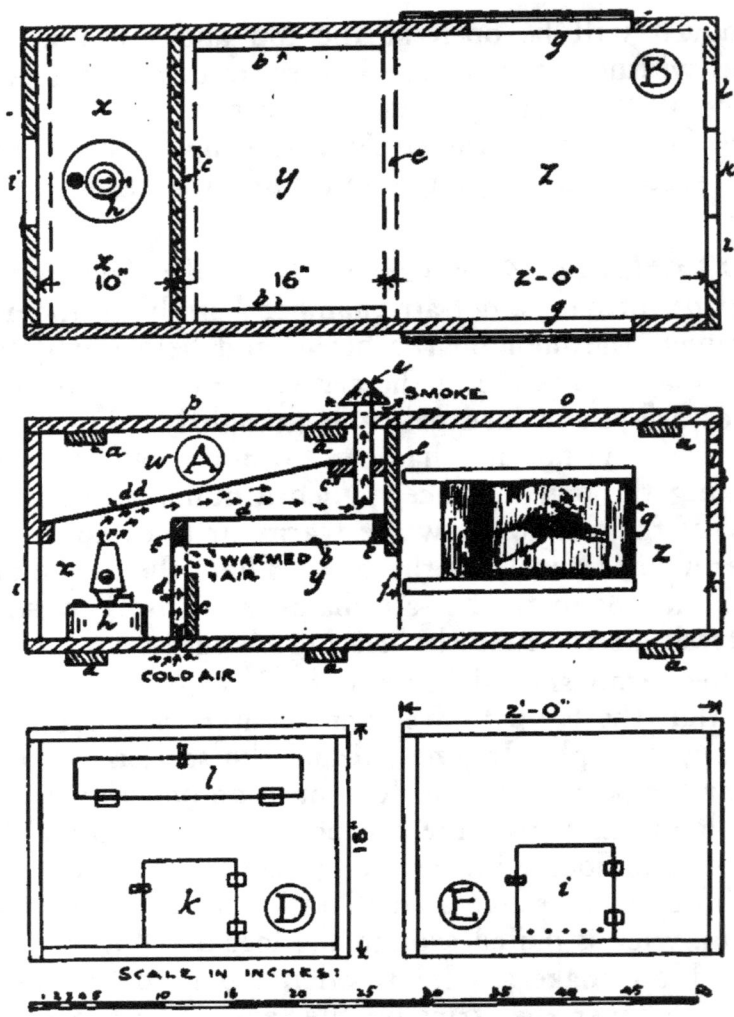

MR. BATCHELDER'S HOT-AIR BROODER

This and the following illustration show the construction. (A) is a cross-section. (B) the ground plan. (C) the top view. (D) elevation of the front end. (E) elevation of the rear end. (F) sectional view of the back of the hover compartment showing the outside air being warmed, and passing into the hover. (G) cross-section of (F).

The abbreviations show: (a) cleats that extend from side to side of the brooder. (b) cleats and boards that are used in making the hover compartment (y). (d) sheet iron extending from the floor of the lamp chamber (x) to the center of the brooder. (dd) deflecting sheet of iron to carry the fumes of the lamp to the smoke pipe (e). (e) smoke pipes. (f) slashed flannel entrance to hover. (g) sliding panes of 8 x 12-inch glass for windows. (h) lamp. (i) lamp door with ventilating holes. (k) chick door. (l) ventilating door. (m) inlet for fresh air in (c) and the floor of the brooder. (n) inlets for warm fresh air in the sheet iron back of the hover. (o) hinged cover. (p) solid cover. (w) tight compartment above lamp chamber. (x) lamp chamber. (y) hover compartment. (z) exercising room. These abbreviations refer to both illustrations of this brooder.

became warmer it could be a fireless. Any mites in the brooder will hide on this cloth. If the top is lifted up and the soiled cloth removed, it can be boiled and the mite family will stop housekeeping.

161. *Dimensions.* A brooder about 3 ft. square or 3 x 4 ft. is used, depending on the size of the room. A hover 2 ft. square or 18 in. by 2 ft. will brood fifty chicks, and that is enough for one brooder — forty is better. For a cover on the brooder while the chicks are young, Mr. Deering takes the balance of the dry goods box — if it is sufficient — and makes the top 2 or 3 in. higher than the hover. The cover is hinged at the back, so that it can be easily raised. We prefer to use unbleached muslin for the cover top nailed on a frame, instead of a top of soiled wood. Two openings with panes of glass to slide on the outside are cut, and screen wire is tacked over the inside. This gives the chicks light and ventilation. There is one opening for a door in front by which a runway to the ground is reached.

162. *Cost.* Mr. Deering is particular to use dry matched lumber, and especially for the hover top and the floor. A No. 2 burner is plenty large, and on some of his brooders he has No. 1 burners. Chicks over three weeks old do not need a cover. They do not require to be confined, as they have learned to return to the brooder when they require warmth. These brooders cost with lamp complete about $3 each. Outing flannel is the best for hover cloth and extra ones are made, so they can be changed every week. Screen wire is tacked around the base to keep the chicks on the floor when the cover is raised. Mr. Deering raised 300 chickens last year and never saw a mite during the season.

163. *An Improvement to Overcome a Trouble.* Mr. Dautrey has improved on Mr. Deering's brooder by adding a second floor 1½ in. above the wooden floor of the heating chamber. This floor is held up by four feet at the corners, and by using it the floor of the brooder is kept cool and comfortable. Brooders made after Mr. Deering's plan should have an inch of garden earth continually on the floor to keep it cooler, otherwise the chicks will suffer from leg weakness. Mr. Dautrey had trouble with his chicks and conceived the idea of the elevated floor; this has given him the best of satisfaction.

164. *Another Plan of Construction.* Mr. Batchelder uses an excellent brooder that does not heat the floor of the brooding com-

partment, and which has the advantage of resting on the ground and not requiring the chicks to ascend to a platform on entering it. In building his brooder it is necessary to construct a box 4 ft. 4 in. long, 2 ft. wide and 18 in. high, outside dimensions. The box is

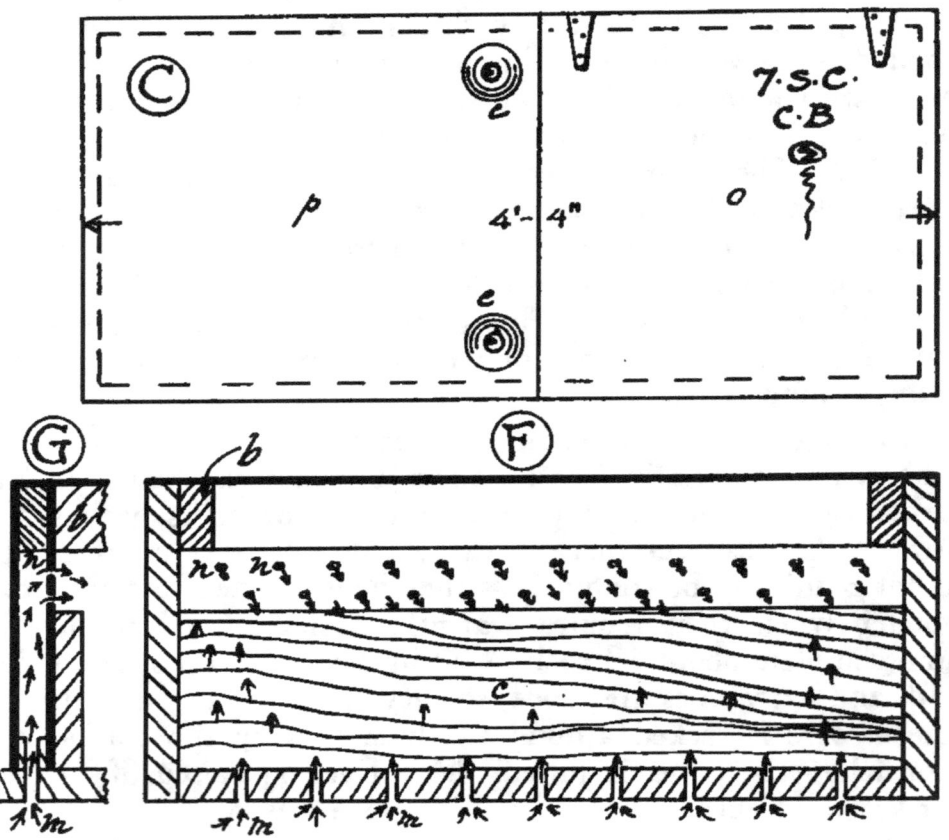

MR. BATCHELDER'S HOT-AIR BROODER
The letters and abbreviations on this illustration are explained
under the previous cut.

built throughout of ⅜-in. planed lumber, or it can be made from packing boxes. As is seen in drawing A, there are six cleats under the top and floor. These extend from one side to the other. The roof is divided 2 ft. from the front end, and the front section should be hinged to one side of the brooder as shown in C. In the front end there is a ventilating door 17 in. long and 3 in. wide; also a chick door 7 in. square. In the rear end there is a lamp door 7½ in. square, with a number of half-inch holes 1 in. above the lower edge. In the sides of the exercising compartment are two

8 x 12-in. windows. These are made of panes of glass that slide in grooved supports, and they can be opened for ventilation. The windows and ventilating door are covered on the inside with 1-in. mesh wire netting to prevent the chicks jumping out when the windows or door are open.

165. *Further Details.* Directly under the opening between the two portions of the roof a $\frac{1}{8}$-in. board c 10 in. wide is nailed. This board extends across the brooder. Ten inches from the inside of the rear end of the brooder, at the floor, a cleat c should now be placed. This cleat is $\frac{1}{8}$ x 1 in., but before it is fastened to the floor a number of $\frac{3}{8}$-in. holes should be bored through it on the $\frac{1}{8}$-in. side. When the cleat is fastened to the floor these $\frac{3}{8}$-in. holes are extended through the floor, so that the outside air can enter the brooder freely through the round openings.

Seven inches below the top edge of the side of the brooder, and above the cleat c we have just fastened to the floor, there is a cleat $\frac{1}{8}$ x 2 in. extending across the brooder. A similar $\frac{1}{8}$ x 2-in. cleat is nailed to the 10-in. center board c at the same level — or with the lower edge 1 in. above the bottom of the center board. Two more $\frac{1}{8}$ x 2-in. cleats b (shown in **A** and **B**) are nailed to the side walls between the two upper cleats cc. On top of these four cleats cc, bb, a piece of galvanized iron should be securely fastened with brads, and on the lamp side of the rear top cleat and the lower cleat with the ventilating holes in it, another piece of galvanized iron should be fastened. We have now the brooder divided by a kind of steps into two compartments.

A $\frac{1}{8}$ x 1$\frac{1}{2}$-in. cleat should be nailed across the rear end of the brooder, with its lower face even with the top of the lamp door. From the upper side of this cleat to another board $\frac{1}{8}$ x 4 in., nailed to the 10-in. center board midway between the galvanized iron and the roof, is fastened the deflecting sheet of iron marked dd in **A**.

The smoke pipes e are made of 1$\frac{1}{2}$-in. tin tubing and reach to within one inch of the galvanized iron sheet d. They are lowered in this manner so that a greater portion of the heat of the lamp will be retained in the heating chamber x.

On the right-hand (front) side of the floor cleat with the holes in it and the cleat c over it, there should be another sheet of galvanized iron fastened. This iron, however, should be perforated

[107]

with two rows of ¼-in. holes near the top. These holes are plainly shown at n in F and G.

It is only necessary for us to nail a slip of flannel f in front of the opening to the hover compartment to complete the brooder. It is better to have two strips, and to cut them alternately every three inches. In this way more heat is retained in the hover compartment.

The lamp should be made with a 7-in. galvanized iron bowl, and one of the old-style hinged burners. A short tin chimney with a mica window is necessary. The total height of the lamp, including the chimney should not be more than 7 inches.

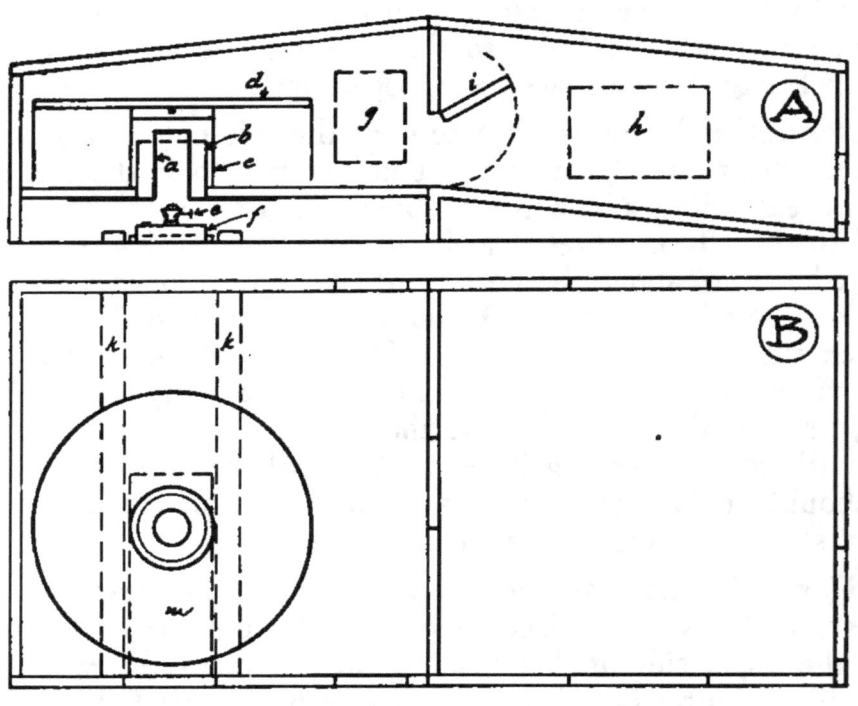

MR. FULLER'S HOT-AIR BROODER

(A) cross-section of brooder. (B) floor plan. (C) drawing of the completed brooder. (D) parts of the hot-air heating system. (E) water fountain for chicks made from a one-gallon bottle and five-cent crock.

Abbreviations for this and the next illustration: (a) air-tight heating drum fastened to sheet iron. (b) short piece of stove pipe. (c) wire-covered chick guard. (d) hover. (e) special burner that requires no chimney. (f) lamp. (g) 6 x 8-inch windows. (h) 8 x 12-inch windows. (i) door to hover compartment. (k) guides for sliding lamp board. (m) lamp board in which lamp sits. (n) ventilators. (o) door to lamp compartment. (p) door to exercising room. (s) hole in top of chick guard in which the pivot of the hover rests.

166. *What the Builder Has Learned.* Mr. Batchelder gives his opinion of this brooder: "I have had very good success with this brooder, and I like it mainly because the chicks are not compelled to ascend a stairway or inclined board, when they enter the brooder and hover compartment. They are constantly supplied with warmed, pure air and there is, therefore, no danger of the hover compartment becoming foul. When the ventilating door 1 in the front of the exercising pen is opened, the warmed pure air enters at the rear of the hover, flows over the backs of the chicks, and passes out of the brooder by the ventilating door in the front. I consider this an ideal system of ventilation and I have been very successful in growing chicks by means of the brooder."

167. *How to "Keep the Floor Low."* Mr. Fuller uses a hot-air brooder with a special lamp and burner that enables him to keep the floor low. A good, tight, dry-goods box 3 ft. square and 3 ft. deep is necessary for building the brooder. Cut in two pieces from end to end, on an angle, so as to have one end of each piece 16 in. deep; the other ends of both pieces will then be 20 in. deep. After removing the end boards from one, place the two deep ends together. Fasten them with four small bolts, so that the brooder can be easily taken apart if it is necessary to store it. This gives us two sections each 3 ft. square. In the section where the hover is to be placed, it is necessary to remove the bottom boards and raise them 4 in. This allows room for the brooder lamp to be placed directly under the heater.

Make a circular hover 24 in. in diameter. This is placed in the corner of the section with the outside edge 1 in. from the end and one side. Cut a circular hole 6 in. in diameter in the floor, directly under the center of the hover. This is to insert the heater from the under side of the floor.

Make, or have your local tinner, a piece of sheet iron 18 in. square, with a hole in the center 3 in. in diameter. Over this 3-in. opening fasten a drum of the same material 6 in. high, with the top end solid and air-tight. This inverted drum is attached to the 18-in. sheet iron by crimping — no solder must be used, as the heat of the lamp would melt it. This 3 x 6-in. drum is inserted through the 6-in. opening of the floor, and extends up in the brooder about 5 in. Fasten the 18-in. sheet iron to the under side of the floor with small, short screws.

On the outside of the heating drum and inside the brooder, place a 6-in. piece of ordinary stove pipe 5 in. high, to direct the heat upwards. Outside of the stove pipe ½ in., there should be a circle of ½-in. mesh stout wire. The wire is tacked to a solid board top. There is a ½-in. hole in the center of the top to insert the pivot of the hover.

A wooden pin at the center of the under side of the hover, and inserted in the wooden top of the chick guard, will keep the hover

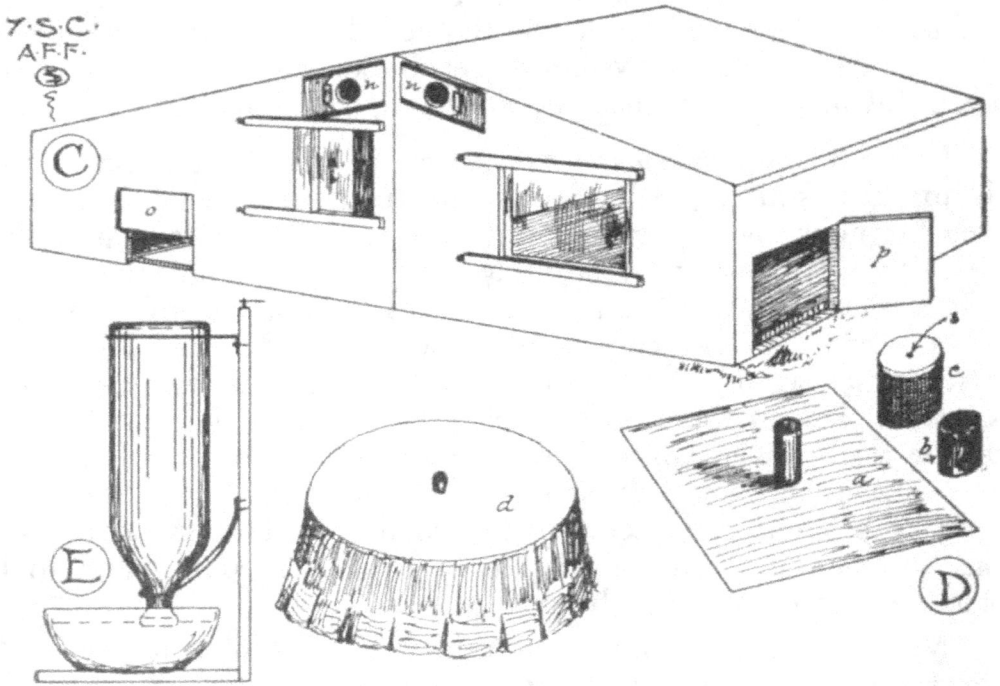

MR. FULLER'S HOT-AIR BROODER
The letters and abbreviations on this illustration are explained under
the previous cut.

in position. Around the edge of the hover tack two strips of old flannel. The lower edge should be slashed every 3 in. (alternately).

Put in two window lights in each side, as shown in the illustration, to let in the light and to serve as ventilators. There should be a door in the partition of the hover compartment to connect with the other, or exercising compartment. In the latter have a small door at the front corner to let the chicks out on the ground. As the floor of the heating compartment is about 4 in. above the ground, a temporary floor slanting down from the center of the brooder to the front is made. This slight slant makes it easy for the chicks to enter the hover compartment.

168. *The Special Lamp.* Mr. Fuller states that a round, galvanized brooder lamp can be bought for 50 cents that is only 3 in. high, and fitted with a burner that needs no flue. This in his estimation, is the best lamp for heating brooders he ever used. A circular hole of sufficient size to allow the bowl of the lamp to rest in is made in a board that will slide under the brooder between two 1 x 2-in. wooden guides. The board is long enough so that the flame of the lamp can be placed directly under the opening in the sheet iron, above which the air-tight drum extends into the hover compartment.

No fumes of the lamp can possibly enter the brooder, and the floor is kept moderately warm. For ventilation four holes 3 in. in diameter are made near the top of the brooder — two in each section — and covered with tin slides. There should be 1-in. holes in the lamp compartment near the ground to supply air to the lamp.

Make a tight board cover, hinged at the center, and cover it with tar paper. If the box is not sufficiently tight, it can be lined inside with any kind of heavy building paper to make it warmer.

CHAPTER XII.

SHIPPING DAY-OLD CHICKS AND EGGS FOR HATCHING

169. *Delighted With This Business.* A short time ago we had the pleasure of visiting Quality Hill Poultry Yards and talking over the day-old chick business with the proprietors Messrs. F. A. and Wm. Z. Bennett. These fanciers are greatly delighted with their day-old chick trade. Mr. F. A. Bennett told us, "I am just carried away with the day-old chick business. We sent out six shipments one day and from four of our customers we received very nice letters. The other two must have been pleased or they would have complained to us." He allowed us to read a number of letters from satisfied customers who had bought day-old chicks and also stated: "Those are the kind of letters we got and not a kick did we receive. Now you know that's wonderful, when you consider the routes and conditions. We advise everyone to buy day-old chicks instead of eggs for hatching, because it is more satisfactory to buyer and seller. For $10, the cost of one sitting of our best eggs, the breeder can get ten chicks hatched from the same eggs, and we guarantee the safe delivery of the chicks within a two days' journey. The buyer runs no chances, as we guarantee that the chicks will arrive in good condition. The chicks are better off without feed the first day or two and our customers have told us later that the chicks they bought are just doing fine."

170. *Mr. Fields' Box.* For shipping day-old chicks Mr. Fields says that a light wooden box 3 in. high is best. This prevents their being jostled up and down by the express clerks, and still does not force them to remain in a cramped position. The latter will often cause deformity. Allow 4 sq. in. of space to each chick. This will give sufficient "elbow room." The box may be made any length and breadth desired, using 4 sq. in. per chick, but 3 in. is the standard height for box. Bore 1-in. holes about 3 in. apart around all four sides of the box. Leave the bottom and top free from holes.

This shipping box should not be made warm by extra cloth, etc., except in extremely cold weather, as the express cars are generally hot enough. However, the boxes should be covered when carrying to and from the station to prevent chilling the chicks. Where a large number of chicks are to be shipped the boxes should be nailed in tiers one above the other each with a capacity of eighteen chicks. (That is, each box a yard square.) As many as a dozen boxes may be so arranged. When tiered, be sure to nail a stout leather handle on each of the two opposite sides.

171. *Directions for Shipping.* Label the box plainly, with instructions *Not to Be Fed and Watered.* The chicks, in emerging from the shell, gather into their bodies enough food and water to last them seventy-two hours. Hence, they will not be real hungry until about the time they reach their destination. Another reason for labeling them thus, is, that the express clerks, in feeding fowls generally dump a great mass of food into the coop. This, in the case of little chicks, would mean death sooner or later. Do not feed the chicks before shipping, but instruct the consignee beforehand to feed and water them as soon as they arrive.

172. *How to Use Baskets.* Mrs. Atwell uses the common cheap baskets, called generally "market basket," and costing about 8 to 10 cents each for the size holding about one-half bushel. They are about 18 in. long, 10 in. wide and 7 in. deep. She selects a good, flat bottom one and puts a mat of excelsior in the bottom, and then lines the inside of the basket with burlap — double *if* early in the season when the weather is cold. A bag such as middlings come in answers very well, when cleaned. This is sewn to the top edge of the basket with a large twine needle, also fastened through the holes at the corners and along the sides at the bottom, so as to keep the lining tight and smooth.

173. *Finishing the Top.* Another piece of burlap is then taken, doubled if the weather is cold, and laid across the top of the basket. It is sewn all around the edge, except a place at one end which is left open to put the chicks in. When they are in, this opening is sewn up also. The burlap cover is not drawn tight like the sides, but allowed to sag a little so the baby chicks can feel it on their backs. This kind of a basket will hold fifty chicks nicely, and they travel much better if they do not have much spare room, as they are not thrown about by the jolting of the train. The burlap sides also help to break the jolts and jars.

174. *Mrs. Atwell Says*: "Their own heat keeps them comfortable. I have used this method of shipping for years and have never had any complaints of dead chicks. I have shipped sometimes when the weather was decidedly uncomfortable for me in driving to the station with them, and I have been almost afraid the 'babies' would

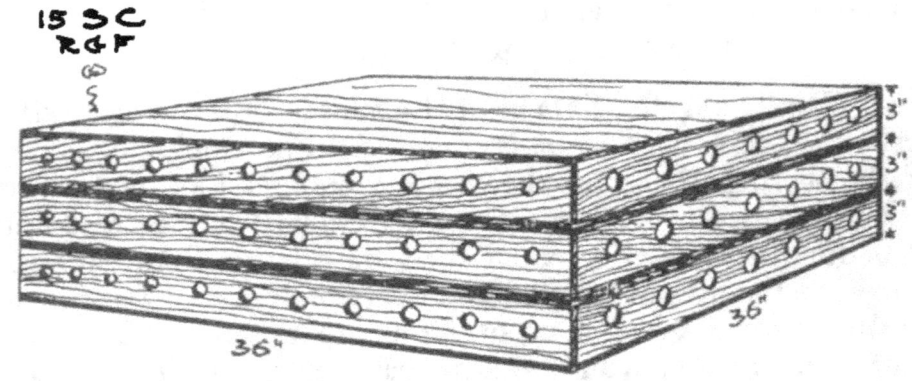

THREE BOXES FOR SHIPPING DAY-OLD CHICKS
Their construction is explained by Mr. Fields.

chill on the way, but my customers always report arrival in good condition. This way of shipping has several recommendations. The package is light, convenient to handle, cheap, easily procured at almost any country store, easily prepared, and lastly and most important, it is comfortable for the chicks."

175. *Another's Experience.* Mr. Vandervort has given this branch of the poultry business careful attention and has the following advice to offer: "A great many persons are very much surprised to hear that chicks only just hatched can be safely sent by express from 200 to 1,000 miles, and reach their destination alive and in good condition. But such is the case, and today thousands of them are shipped during the hatching season to all parts of the country. Some people do not know that a chick, which escapes from the shell in which it has been a prisoner for twenty-one days, will stand a lot of abuse and live, provided you are not too harsh with it. How much feed and care a little chick needs is a question open for discussion. Nature has provided a supply of nourishment for the little fellow in the form of the yolk of the egg from which it it hatched. As the chick comes from the shell with the undigested yolk yet in its body, it furnishes all the food that the chick needs for the first seventy-two hours of its life.

"I have given this matter thorough thought and study, and have come to the conclusion that when the little chicks are packed away for several hours during shipment they continue to gain strength and are much better off than if they were exposed to the sudden changes of running in and out from the hover of the brooder. And this is why so many are safely shipped, and reach their destination without losses and in good condition.

"After the chicks are hatched and nicely dried off, they are ready for shipment. Baskets are excellent for shipping the little fellows in small lots of from twelve to fifty. These baskets should be strong and low, about four inches is the proper height. Line the basket well with burlap, and cut clover and chaff should be placed in the bottom. Put your chicks in and sew a cover of burlap over the top of the basket, put a label *Live Chicks* in a conspicuous place, and they are sure to reach their destination in good condition. I like baskets best for they are light, and easily handled. Any manufacturer can supply you with any size basket you wish.

"If you prefer to ship in boxes make them of light material and 4 in. high, and they should be lined the way mentioned for the baskets. If any great number are to be shipped, the chicks can be placed in boxes containing several decks. Don't feed them anything for their journey, as it is injurious to them. The shipping of day-old chicks has come to stay. Buying young chicks will save you a lot of trouble and money, too. Send your orders early and do not expect to have your chicks shipped the day after you order them, for you know you cannot get them until after they are hatched and there may be a lot ahead of you."

176. *A Box With Two Compartments.* Mr. Roberts has had good success by using a home-made box as follows: He makes a box 14 x 8 x 4 in. deep of ¼-in. sheeting (pine). Heavy card or strawboard is used for the bottom, and the top is covered with burlap over which is tacked strips of the sheeting 1 in. apart. The inside is lined with flannel. The box is divided at the center, making two compartments 8 x 7 in. These will accommodate from twelve to fifteen chicks on each side, allowing about 4 sq. in, to each chick. When cardboard is used for the bottom, two strips of the sheeting should be nailed lengthwise outside as a protection. Just before placing the chicks in the box, Mr. Roberts puts a handful of feathers or fine shavings in each side.

177. *How to Ship Eggs.* Eggs for hatching should be securely packed, but yet in a light package so as to eliminate heavy express charges. There are several ways in which to ship eggs for hatching safely. One is a patent egg basket which seals itself when closed by the handle, being made of corrugated pasteboard and divided

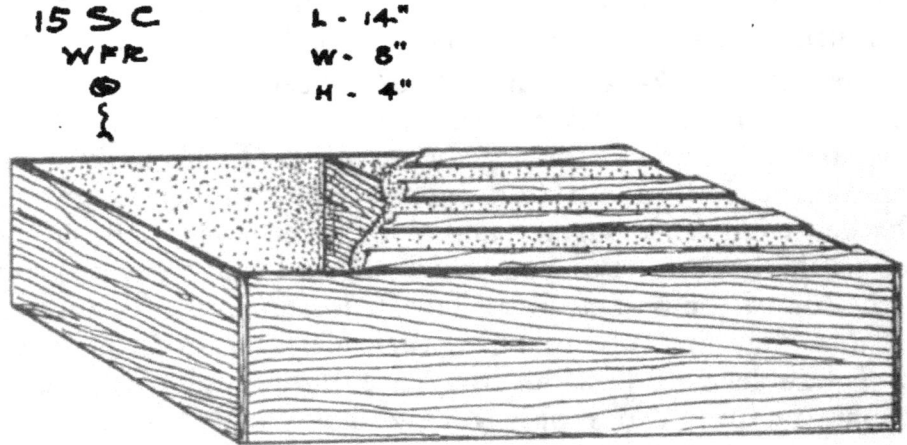

MR. ROBERTS' BOX FOR DAY-OLD CHICKS

into fifteen compartments. When using these egg carriers always wrap the eggs in soft paper, using sufficient paper in wrapping to make each egg fit tight in its compartment, and thus prevent rocking.

178. *Advice on this Subject.* Mr. Albeck gives the following information: In placing eggs in the basket, always place them with the *small* end down. The air cell of an egg is in the large end, and in shipping may become displaced by rough handling, or by the weight of the contents being forced upon the tissues that hold it in place; thus the egg becomes addled, killing the germ. By placing it on the small end this is overcome.

Also be sure and notify the buyer to unpack and lay the eggs natural for at least twenty-four hours before setting, so the eggs can cool and take their natural shape. This will result in a much better percentage of chicks from the eggs.

While these egg carriers are easier packed than common market baskets, the baskets are cheaper, will hold more than one sitting, and can be packed for cold weather. Another advantage is that one does not need to carry a stock on hand as they can be bought at any grocery store. In packing these baskets, first line the bottom and sides with good, clean, heavy paper over which place about 2 in. of soft excelsior. In wrapping each egg, have soft paper about 8 x 12 in. in size, laying the egg on one corner and give two wraps, then fold ends in and finish wrapping.

After the eggs are wrapped take soft excelsior and wrap each egg over the paper, and place in the basket small end down. Now take more excelsior and pack sides and small spaces between eggs firmly so every egg is immovable. If more than one sitting is to be packed, cover the eggs with about 2 in. of excelsior, over which lay a piece of pasteboard to make a foundation for the next layer, packing eggs same as the bottom layer. After the eggs have been securely packed, put plenty of excelsior on top, over which sew a piece of muslin.

In putting on the muslin use a blunt knife, or putty knife, and tuck the ends between the top rim. Now take a large darning needle and strong twine, and sew through the basket and over outer rim. Print on muslin, or use gummed labels, *Eggs for Hatching*. As these baskets generally have two handles, they should be tied together as it prevents setting other express packages on top, and also keeps handles in place for handling.

To the handle there should be a tag stating plainly the consignee's express office, also postoffice address. Since so many rural routes have been established in the United States, a number of the small towns that receive express from railroads have a different postoffice address, as a rural route goes out of a larger city to their village. When done in this manner, as soon as the eggs arrive at the express office and consignee is not there to receive them, the express agent can notify him that package is there and save much delay.

179. *A Breeder's Views.* Mr. Vandervort gives his experience in shipping eggs thus: "It requires considerable skill and a good knowledge of details, to pack eggs for hatching so they will reach their destination in good condition and hatch well. Some use boxes, and while they may be all right, I have had much better results with baskets. Eggs shipped in boxes are liable to be thrown around and roughly handled, while baskets will be handled much more carefully, and are much cheaper. Use baskets with a strong upright handle, so as to guard the package from other matter being laid upon it.

"I exercise the same care and follow the same rule for packing one sitting as we do for one hundred eggs, except as to the size of the basket; always having our package just large enough to hold the eggs and the necessary packing material. If the order calls for 100 or 120 eggs, a bushel basket is needed.

"First, line the basket with newspapers and then put about one inch of excelsior in the bottom. Now we are ready for the first layer of eggs, which, after wrapping each egg in soft paper, are laid singly, leaving a 1-in. space between the eggs and the sides of the basket; also a $\frac{1}{2}$-in. space between each of the eggs. These spaces are filled with excelsior, crowded in firmly so as to keep the eggs not only from the sides of the basket, but to keep them from shaking.

"This bottom layer of eggs is then covered with 1 inch of excelsior in the same manner, and all the other layers likewise packed and covered as the others, with the exception of the top and last layer covering, which should be thicker and rounded up so that when the canvas cover is sewed on and drawn down tight, the eggs cannot by any possibility shake around in the basket. The label can best be sewed to the canvas cover before the cover is put on the basket.

"Have *Eggs for Hatching — Handle with Care* printed on your label in good size type. Use a darning needle and strong cotton string and sew the canvas to the rim of the basket, drawing it down tight, so that the packing will spring up when pressed with the hand.

"A basket about the size of a ten-pound grape basket will hold a sitting of eggs. For two sittings, a one-fourth bushel basket is necessary, and for fifty eggs a one-half bushel basket is required. Excelsior for packing the eggs can usually be procured at all grocery or furniture stores. For wrapping the eggs, a soft grade of newspaper is used. The eggs are placed in the baskets on end, preferably with the little end down."

180. *Most Suitable Eggs.* The following general information by Mr. Vandervort about the kind of eggs that are most suitable for shipping, and that will produce the most chicks is interesting: "Eggs to grow strong, healthy chicks must be fertile and laid by healthy hens I believe fertile eggs even from healthy fowls, are

largely a question of the right ration. Therefore, in the winter we try to supply our breeders with about one-fourth ounce of cut bone daily, and for the grain ration at least one-half is oats.

"Fertile eggs to ship safely must have strong shells. For this purpose keep oyster shells before the hens all the time, and plenty of green food. I have heard complaints of shells of eggs being too thick to hatch well in an incubator, when the hens were allowed all the oyster shells they could eat, but our trouble has been to get the shells strong enough, especially when the flock averages five eggs for each hen per week.

"It is possible that a hen that lays only every other day, or one whose shell machinery is extra good, might lay eggs with too thick shells. I would risk this danger, anyway, believing that a chicken that has not strength enough to pop open his shell has not vitality enough to make a good, healthy fowl. Fertile eggs with strong shells, rightly packed, properly labeled and from the right kind of stock, can be shipped safely by express, and can be depended upon to hatch well on arrival at the destination."

CHAPTER XIII.

FOOD HOPPERS AND FOUNTAINS

181. *Dry Feeding Gains in Popularity.* Within the last few years the system of feeding all ground grain mashes *dry* in hoppers instead of moistened with water or skim milk has had a wonderful gain in popularity. In the first place, hopper feeding saves considerable of the poultryman's time, and in the second place it is a more natural way for hens to take their food. The primary disadvantage of feeding wet mashes is that the hens fill their crops with a wet food that is in the proper condition to decompose — unless it is quickly passed on and digested. "Sour crop" is a common complaint of wet-mash fed hens, and it is a trouble or disease that the dry-mash fed hens are never subject to. A big crop full of wet mash makes the bird lazy, and it is natural for it to sit in a corner of the house until the meal is digested.

182. *Hopper-Fed Hens are Better Layers.* Feeders have tried to eliminate the difficulties in feeding wet mash by feeding it in limited quantities, but this is somewhat of a disadvantage because it is possible to supply a balanced ration that will tend to increase egg production through a mash, more easily than in the whole grain. When ground mash is always kept in a hopper the hens will not overeat, and if the mash contains animal food and vegetable matter in addition to the grains, these hens will lay better than other hens that are fed a wet mash once a day.

183. *A Stovepipe Hopper.* Mr. Batchelder gives the construction of his stovepipe hopper that works well. It will not clog and can be easily made from one length of pipe and a few pieces of lumber. Two holes are made near one end, on opposite sides of the pipe, and wires are fastened through these holes to a ring, as shown in the illustration. Another wire or strong cord reaches from the ring to the ceiling of the poultry house.

184. *Construction.* The wooden box has the floor about 4 in. above the ground, instead of right on the ground, so that there will be less danger of the fowls kicking straw and dirt into the food. Wooden strips $\frac{3}{8}$ in. thick by 1 in. wide are nailed on the four top edges of the box and extend inside, in order to prevent the fowls wasting any food.

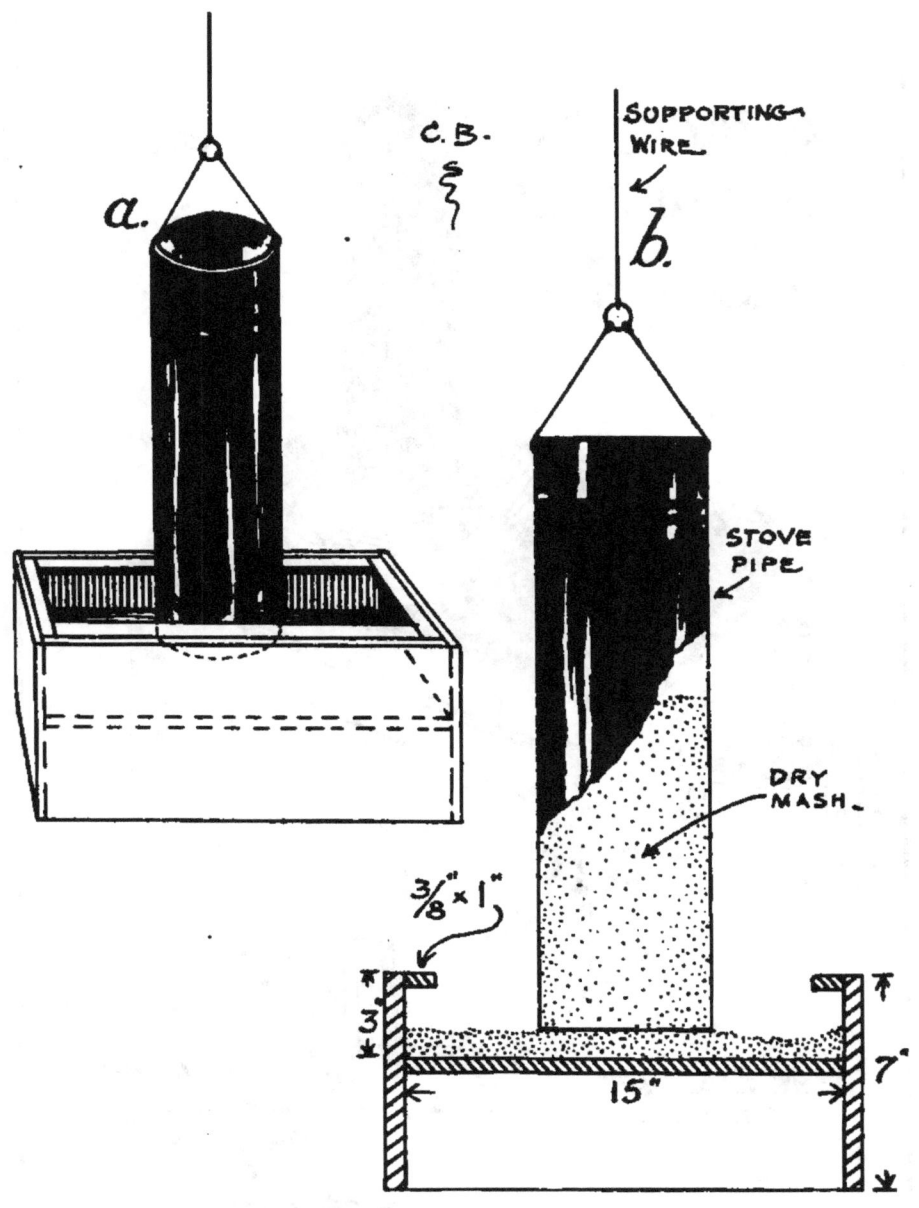

A STOVE PIPE HOPPER

(a) Elevation of hopper showing the stove pipe supported from the ceiling.
(b) Cross-section of hopper giving the principal dimensions.

185. *Adjusting the Pipe.* The bottom edge of the stovepipe is about 1 in. above the floor of the box. If it is found that the feed flows too freely, the stovepipe should be lowered a trifle, or if the

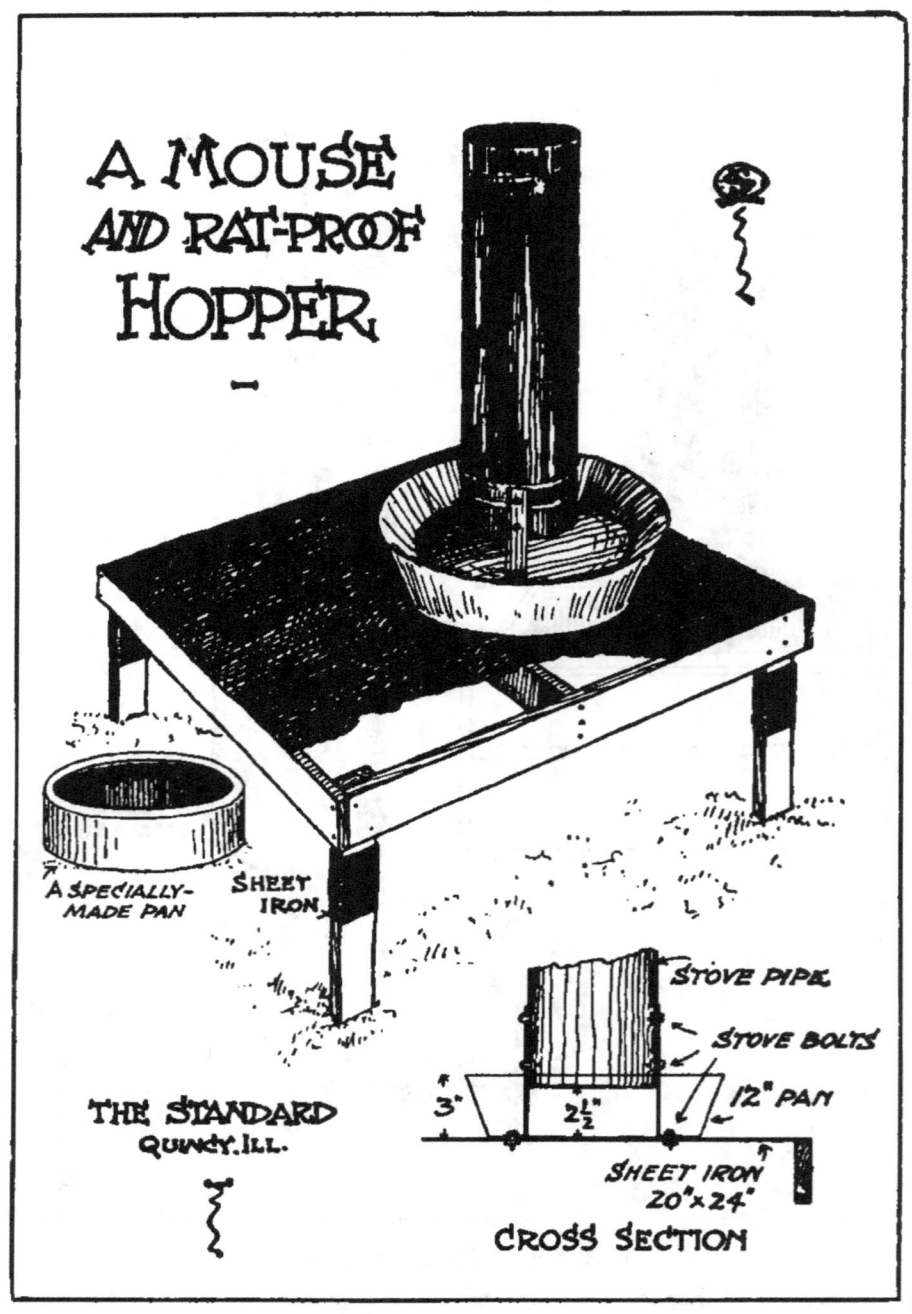

A HOPPER THAT SAVES THE FOOD

feed does not keep the bottom of the box covered the pipe should be raised. The pipe is placed in the proper position and then securely fastened by the cord or wire from the ceiling.

186. *Don't Feed Mice and Rats.* Some fanciers would use a hopper for feeding dry mashes, if it were not for the loss of grain through mice and rats. Only a short time ago we were told by a

A DOUBLE DRY-MASH HOPPER

Mr. Hildebrand made the above hopper from a Columbia oats box. (A) cross-section with the principal dimensions. (B) completed hopper.

fancier that he caught two or three mice in the feeding thorugh of his hopper when he entered his poultry house one early morning. The state of affairs is annoying, even if we do not consider the unnecessary loss of grain and the pollution of it.

187. *How the Hopper is Made.* The accompanying illustration shows a frame made of 1 x 2-in. lumber with a 20 x 24-in. sheet iron top. The legs are 12 in. high and you will notice that a band of sheet iron about 3 in. wide is around each leg, near where the boards forming the frame are attached. The height of the iron top makes it impossible for rats or mice to jump up on it, and they cannot crawl up the legs, because the sheet iron band will not give them any footing.

The hopper is similar to the one used and previously described by Mr. Batchelder. It is made from one length of stove pipe, a common or special feed pan, two 1 x 6-in. pieces of heavy sheet iron and six stove bolts. The 1 x 6-in. supporting pieces should be bent at a right angle, one inch from one end. This makes each piece L-shaped.

[123]

188. *Assembling It.* The hopper is put together as follows: The feeding pan is placed near one end of the stand, and the stove pipe stood upright in it. The two L pieces are set at opposite sides of the stove pipe, and with an awl or nail a hole is punched through the L piece, pan and sheet iron floor. Now, insert a stove

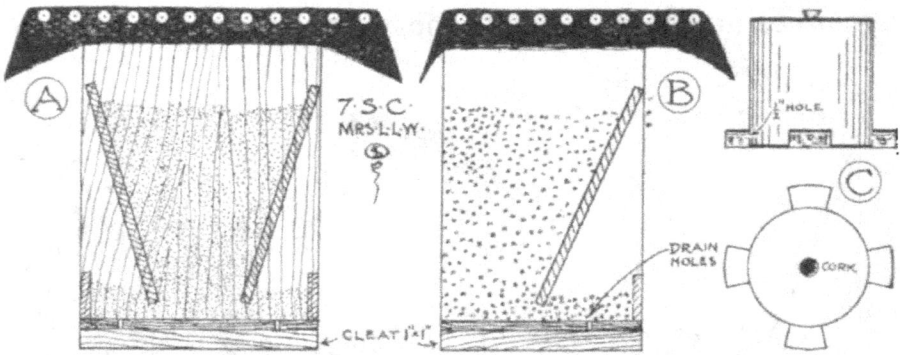

MRS. WHITE'S HOPPERS

(A) hopper for dry mash from which twelve fowls can eat on each side. (B) hopper for grit. (C) water fountain made from a large round can, with four cups soldered to the side.

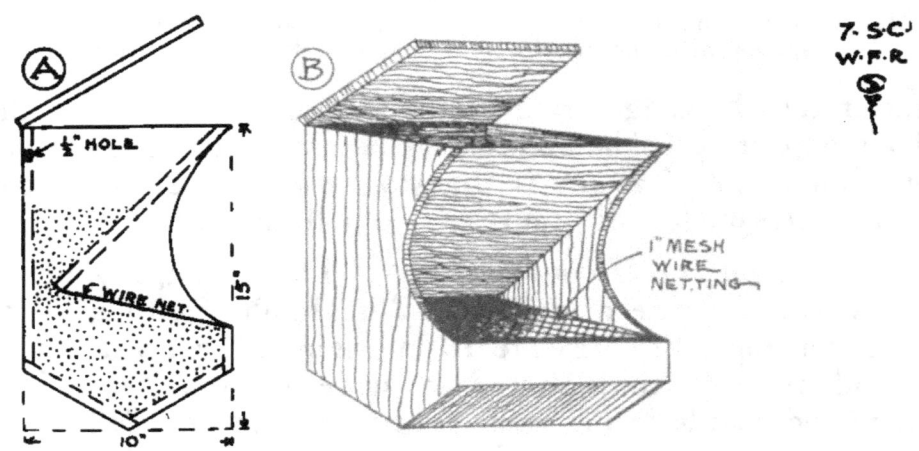

MR. ROBERTS' FOOD HOPPER

bolt in the hole and fasten all together. Punch two holes in the long end of the L pieces and bolt them to the stove pipe. The stove pipe is raised $2\frac{1}{2}$ in. above the floor, when a 3-in. feeding pan is used.

By bolting the parts together, it is impossible for the chicks to stand on the edge of the pan and upset it, which they would probably do if it simply rested on the top of the frame. The fowls jump up on the iron top when they eat from the hopper. It is sufficiently large to hold several at one time.

The stove-pipe form of hopper is the simplest type we are acquainted with, and there is absolutely no trouble with it. Some of the hoppers with small feeding mouths give considerable trouble, on account of the grain becoming packed in the small opening. This effectually prevents the hopper from feeding, until the poultryman takes a stick and clears the obstruction.

189. *A Special Pan.* At the left-hand side of the illustration another feeding pan is shown. This is a specially-made pan 12 in. in diameter and 3 in. high. Its main feature is an inside 1-in. rim around the top. This rim prevents the fowls from drawing grain over the sides of the pan. However, if you select a pan with as little flare as possible, we hardly think you will have any trouble in this respect.

The details of this mouse and rat-proof hopper will prove suggestive to poultry raisers who use hoppers of other styles — from which the mice obtain their living—in illustrating a simple method of overcoming this loss. Raise your hopper from the floor, or if it is attached to the wall, nail a strip of sheet iron on each side, so the mice cannot crawl to it.

190. *From a Packing Box.* Mr. Hildebrand makes an automatic food hopper from a Columbia oats box and it works well. This hopper can be used outdoors or indoors. The dimensions are all given on the cross-sectional view, so that it is not necessary to refer to the construction.

191. *Other Styles.* Mrs. White illustrates the two styles of hoppers which she has found most suitable. The long, double hoppers (A) are made from wholesale shoe boxes, or milliner's hat boxes, and allow about twenty-four fowls to eat at a time. The smaller grit hoppers are made from boxes about 20 in. long and 12 in. wide.

192. *Construction of Mrs. White's.* The construction of the double hopper is thus described. The top and two sides of the big box is removed, two solid boards slanting from the top to the bottom are nailed in, leaving a space at the bottom of about 1 in. for the mash to spill out as the hens eat it. At each end of the hopper

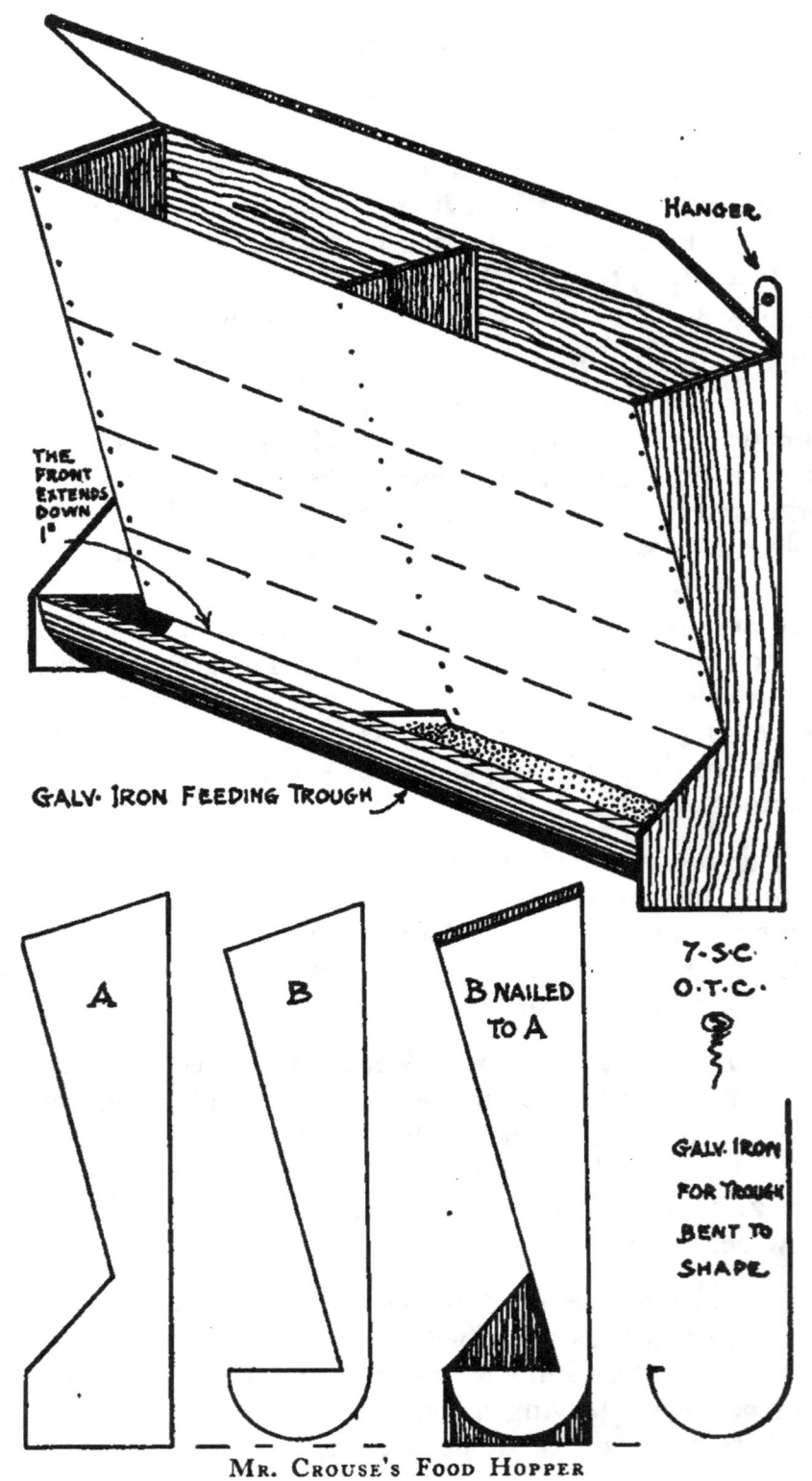

HANGER

THE FRONT EXTENDS DOWN 1"

GALV· IRON FEEDING TROUGH

A

B

B NAILED TO A

7·S·C· O·T·C·

GALV· IRON FOR TROUGH BENT TO SHAPE

MR. CROUSE'S FOOD HOPPER

a 1-in. square strip is nailed on the bottom to raise the box off the ground, thereby lengthening the life of the hopper, and also giving drainage for any water that may have entered from a heavy, driving rain. In the bottom eight or ten holes are bored with a gimlet for the water to leak out. The tops are covered with Ruberoid or Amatite roofing, and the roofing allowed to drop over the side and extend an inch or so over the end. The tops are removable. For at least ten months of the year the hoppers are out in the weather, only being put in when winter weather or snow keeps the fowls housed.

193. *An Original Hopper.* Mr. Roberts has found that his home-made hopper gives better satisfaction than any hopper he can buy. He bought an empty box measuring 15 x 15 x 10 in. at the grocery store. The box is stood on edge and the front taken off. Now insert it on an angle of about 45 degrees from the front top edge of the box, leaving a space of about 1 in. between the bottom edge of this slanting piece and the back of the box for the grain to sift down. In our opinion, it would be preferable to have the mouth of the hopper 2 in. instead of 1 in. There would then be no danger of the hopper becoming plugged with grain. A piece of 1-in. mesh poultry netting is attached to the bottom edge of the slanting front. The netting extends to the front of the hopper and lays on the grain. This prevents waste and is a good idea. Through the back near the top two $\frac{1}{2}$-in. holes are bored, by which the hopper is hung on the wall.

194. *"The Best by Test."* Mr. Crouse says that a satisfactory grit or food hopper was a very difficult appliance for him to obtain, and he tried many of the ready-made, as well as home-made ones. After considerable experimenting he adopted the form of hopper shown in the sketch as being the most practical.

To construct, first make two pieces like A in sketch and three pieces like B. Nail one B to side of A, letting the top down about $\frac{1}{2}$ in. for the lid, and keeping the front edges flush. If the back edges are not flush, they can be made so with a few strokes of the plane. One end is now complete. Construct the other end in a similar manner, being careful to place A and B together so that the two pieces marked B will be on the inside when the hopper is put together.

The other piece B is for the center division of the hopper, to separate it into two compartments, and to make it stronger. Now take a sheet of tin and turn about ½ in. on the long side until it is

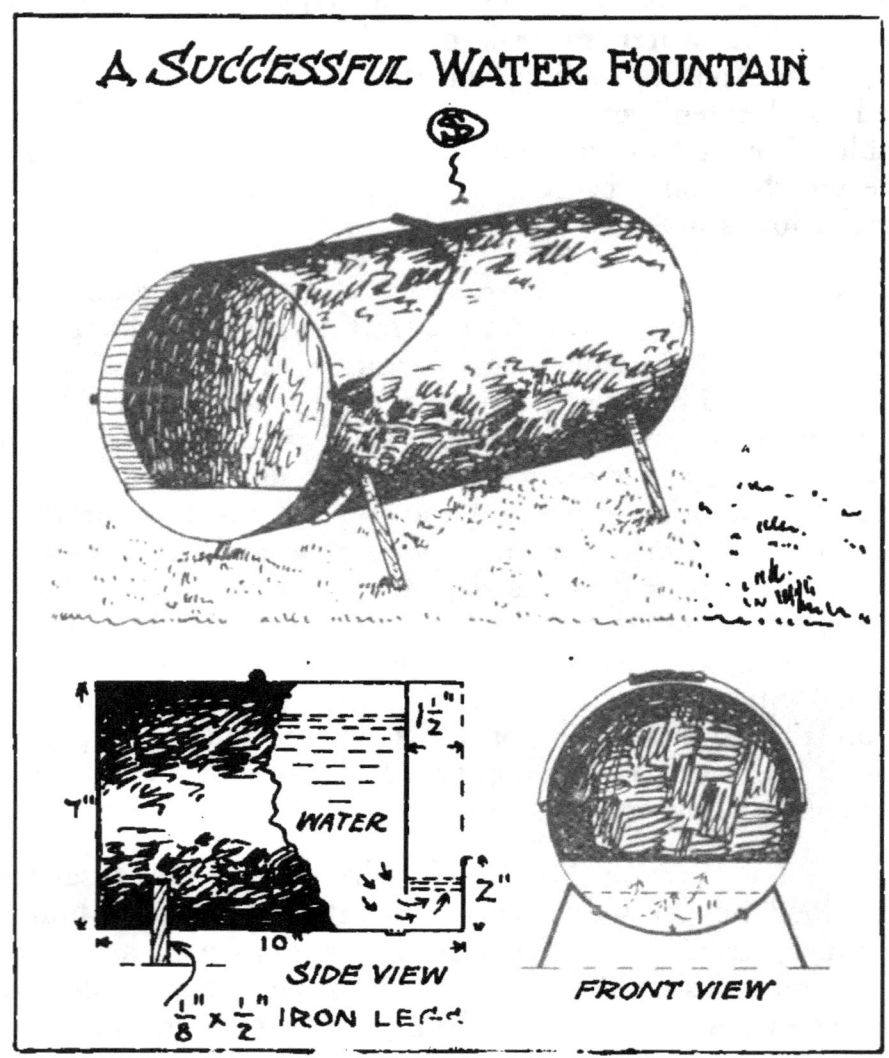

A CONVENIENT AND INEXPENSIVE WATER FOUNTAIN

at a right angle with remainder of the sheet. This part is to extend over the trough to prevent the food from being wasted.

The tin should then be bent so that it will fit around the curved pieces B at the bottom. Place the ½-in. side over the top of the feed trough, and nail the tin to the two end pieces. Then insert the center piece. Complete the back to the top of the end pieces with thin boards. Nail thin boards over the front, beginning at the bottom, and for ordinary dry mash it is best to allow the front to extend down about 1 in. as shown in the sketch.

Mr. Crouse puts a thin strip of wood about $\frac{3}{4}$ in. wide under the tin projection at the front edge of the feed trough. This stiffens the tin and prevents the birds cutting their wattles in feeding. The wood and tin should be smoothed with a file and sand paper. Fasten the lid with small hinges and attach hooks to the back to hang the hopper up, and it is complete.

195. *Water Fountain.* A convenient water fountain can be made from a galvanized iron pail 7 in. in diameter and 10 in. deep. There is no flare to the pail, but it has a wire pail handle at the top. One and one-half inches inside the top a galvanized iron head is soldered. This head is 7 in. in diameter, but it has a 1-in. piece cut off the lower side so that the fountain can be filled and operated. Before the head is soldered, two pieces of $\frac{1}{8}$ x $\frac{1}{2}$-in. band iron should be bent to form the legs; these pieces are riveted to the pail. A lip 2 in. deep holds the water in place. This is an inexpensive fountain, and to fill, it is only necessary to stand it upright and pump water into the mouth. The fountain can be cleaned by throwing a small handful of salt and sand into it, adding water and shaking the mixture around inside.

CHAPTER XIV.

METHODS OF SPROUTING OATS

196. *Makaing Palatable Green Food.* The following excellent system of sprouting oats for poultry is used by Mr. Bennett of Quality Hill Poultry Yards. As he told us, "it is a fine thing." The green food problem presents difficulties to the poultryman during the winter time. It is frequently difficult to buy cut or ground alfalfa or clover hay, and buying vegetables in small quantities is expensive. By means of the system of sprouted oats which

A SPROUTING RACK FOR OATS

Front view showing the seven trays in position.

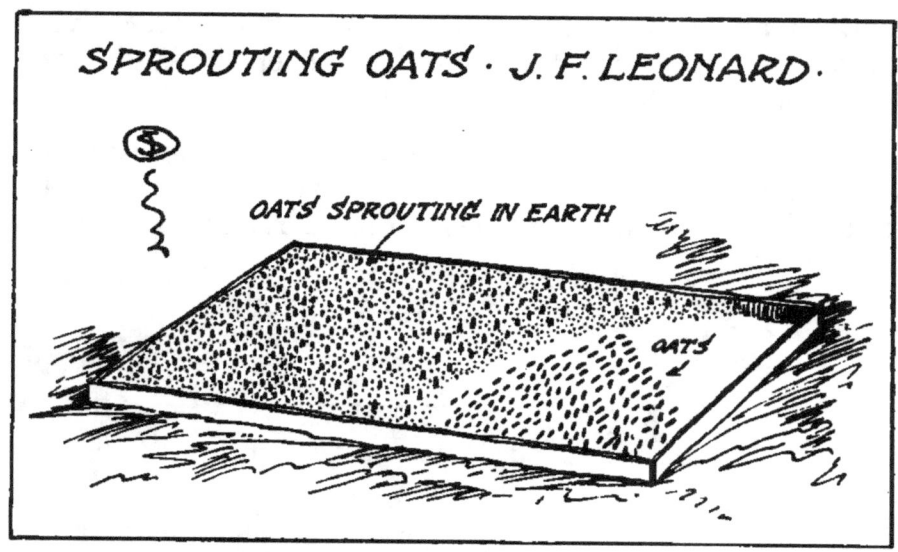

A New Method of Sprouting Oats

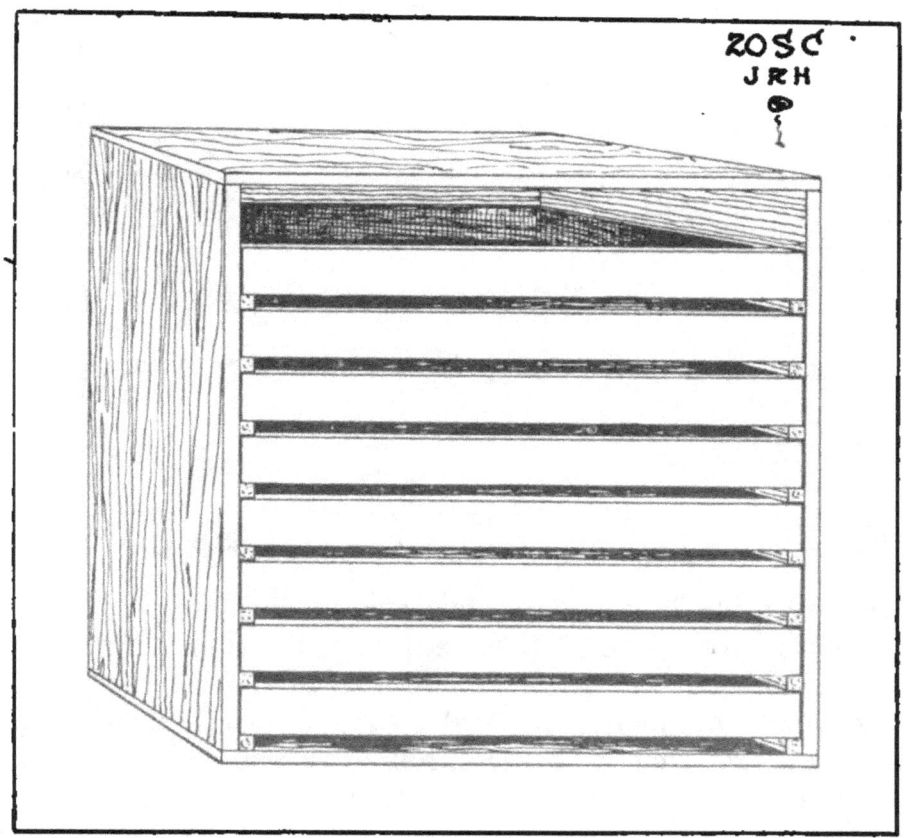

An Inexpensive Sprouting Box
This box is used in Mr. Harris' method of sprouting oats for poultry food.

is described and illustrated, it is a simple and inexpensive matter for anyone to supply his fowls in the winter time with an abundance of succulent food they will relish.

197. *The Value of the Rack*. The simplest way to sprout oats is to put them in a dish in the sink, cover with warm water at night and feed them the next day. This arrangement not only in-

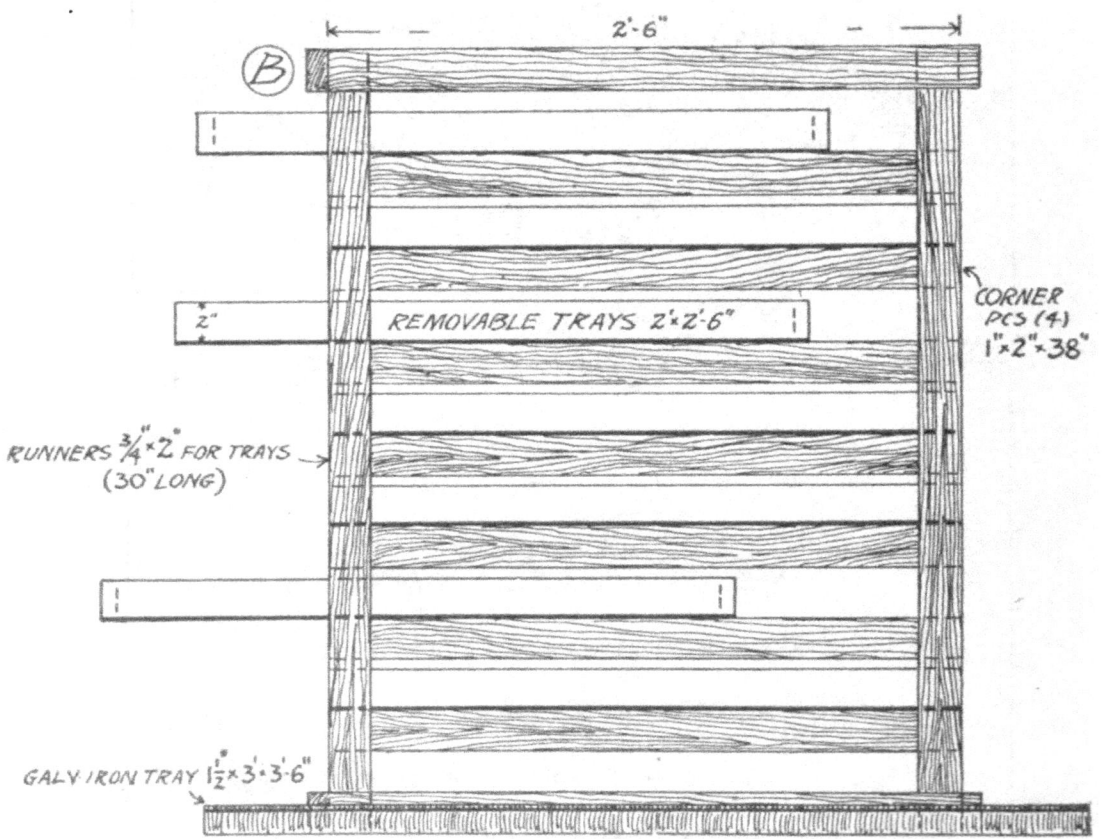

A SPROUTING RACK FOR OATS
Side view showing part of the trays in the rack and three partly removed.

terferes with the house work, but the sprouts are not sufficiently long, where green food is the main consideration. The rack that is shown holds seven trays — one for each day of the week — so that each tray of oats is allowed to sprout seven days before the grain is fed. At that time the sprouts are about 1½ in. long, and much better for green food than oats with sprouts only ¼ to ½ in. long. It may be true that the oats with the long sprouts are not as nourishing as the oats with the short sprouts, but that does not make any material difference when the oats are desired for *green* food and not for *grain* food.

198. *Making the Rack.* The sprouting rack is a light frame made of 1 x 2-in. pieces, without covered top, bottom or sides.

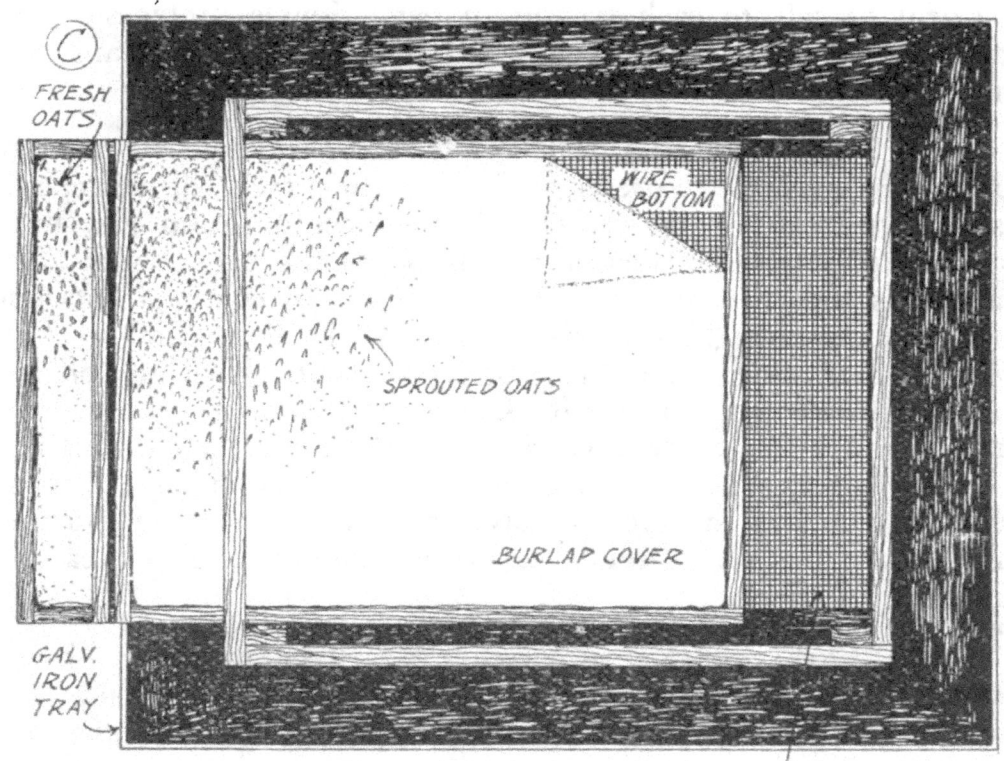

A SPROUTING RACK FOR OATS

Top view. The upper tray is filled with sprouted oats, and the one farthest extended has fresh oats in it. The black border around the rack represents the galvanized iron tray in which the rack stands.

There are only four corner pieces and four pieces at the top and four at the bottom to hold the frame together. On each of two sides of this frame seven runners are nailed 2½ in. apart. On these fourteen runners rest the seven trays. The floor of the tray is covered with ¼-in. mesh wire cloth such as is used for cellar windows. The wooden sides are made of pieces ¾ in. thick by 2 in. wide. The rack stands in a galvanized iron tray which catches the excess water.

199. *Sprouting the Oats.* The trays are covered with dry oats, and hot water is sprinkled over each tray—using a hand sprinkling

can. The trays are then replaced in the rack. They are covered with burlap to keep the oats dark and moist. The trays should be drawn out morning and night and sprinkled with hot water. The rack should be kept in a temperature of from 60 to 80 degrees. It is advisable to season the sprouted oats with salt — using a teaspoonful of salt to each trayful of oats. Fowls prefer sprouted oats to any other green food, and chicks a week old are very fond of them.

200. *The Outdoor Method.* Mr. Leonard has a system of sprouting oats that we prefer to the sprouting rack when it can be used. The new way is less trouble than the old, and the chicks will obtain bugs and worms in addition to the oat sprouts. The method is thus explained. Take one bushel of common oats and soak them over night. Make a frame of 1 x 6-in. lumber, 3 ft. wide by 8 ft. long. Place the frame on smooth, hard ground and spread the oats evenly inside it. Cover the oats with 1 in. of loose ground and water every day. When the sprouts show through, it is ready to feed. Then with a garden hoe, work under the roots, pull them up straight and you have the finest green food you ever saw — food that you cannot beat with anything else as cheap, or as simple to obtain.

201. *"Bugs and Worms" Also.* Mr. Leonard says: "Every hoeful has a big lump of dirt on it, also some animal food. Examine some of this earth and you will find it full of bugs and worms. This gives the birds exercise to scratch them out. Green food grown in earth has more strength and substance than when it is sprouted with water in a warm room. I keep three beds going all the time and have raised this season over 400 chicks to broiler size in four weeks. We ate a pullet three months old the other day and she had already started an egg bag, and I think that is going some."

202. *Sprouted Oats Solve the Problem.* In reference to the value of sprouted oats Mr. Harris says: "For the fanciers that live on farms it is no trial to secure green stuff for their flocks, but when it comes to the city fancier, there is a great deal to contend with. I believe all will agree that hens do not have as satisfactory laying record where they do not obtain an abundance of green food. If the city-lot fancier expects the results his fowls should

bring, he must try in some way to overcome this trouble. A sprouting box is, I think, the best and probably the cheapest way to supply green food."

203. *Mr. Harris' Plan.* A store box 3 ft. wide, 4 ft. long and 3 ft. deep answers the purpose very well. It has drawers or shelves with from ½ to 1 in. of oats sprinkled thereon. The oats should be wet with lukewarm water, and the box placed in a warm room. Leave the oats in the trays until the sprouts are 3 or 4 in. long; then take sprouts and oats and feed all to the chickens. This makes a good noon meal. The chickens will eat it up greedily and will thrive. It will take from a half to a peck of oats to fill the trays. Have it arranged so there will be sufficient oats for each day's feeding.

CHAPTER XV.

OTHER USEFUL DEVICES

204. *A Hook for Catching Fowls.* Some poultrymen use a hook for catching their fowls, and if it is used quietly will often prove a time saver. Such a hook is illustrated. It is made of a piece of $\frac{3}{16}$-in. wire about 15 in. long, and bent as shown. The

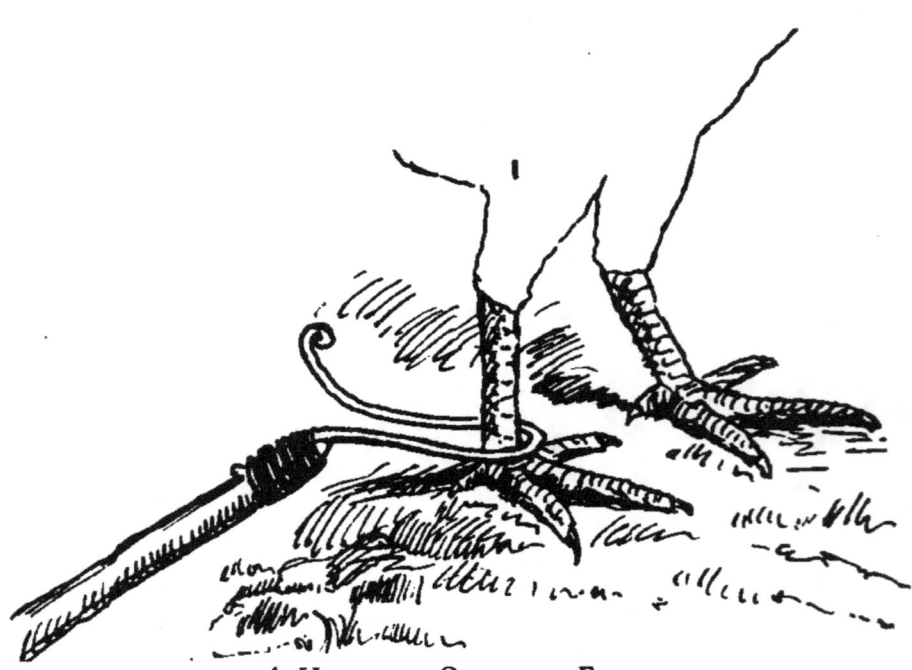

A HOOK FOR CATCHING FOWLS

The upper illustration is a side view of the hook and part of the handle.

hook is fastened by wire to a fishing pole 8 or 10 ft. long for a handle.

205. *Candy Pail Furniture.* Candy pails can usually be bought at low cost at the stores. From these, excellent drinking fountains and nests can be made after Mr. Cavanaugh's patterns. The foun-

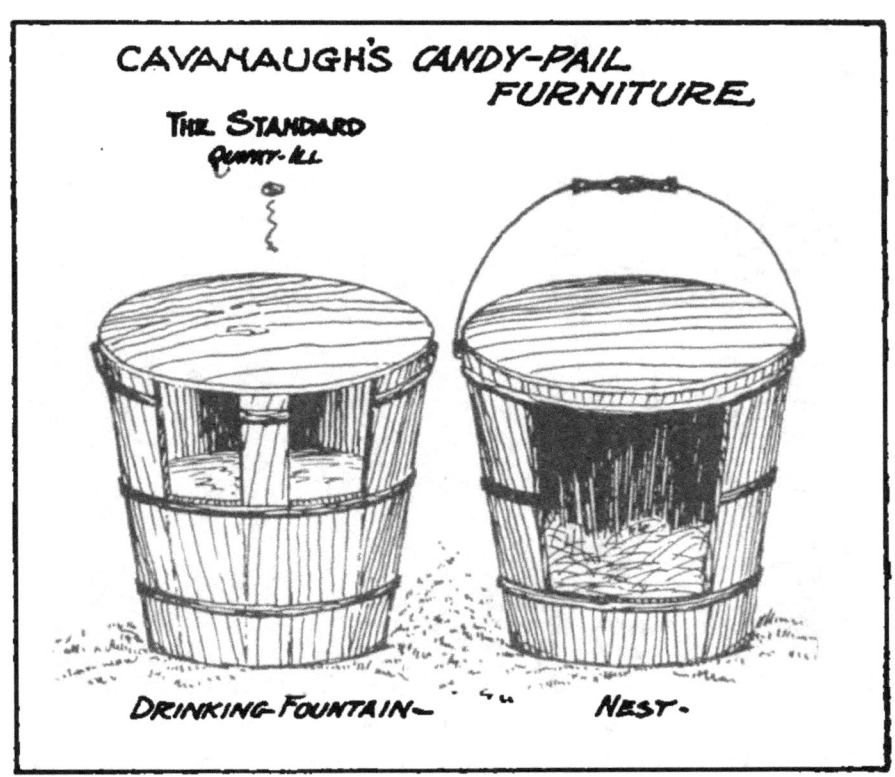

USEFUL CANDY-PAIL FURNITURE

tain has the upper parts of two staves (about 4 x 5-in. openings) removed. The nest has a couple of the staves cut away as shown. The covers are left on, which keeps the water clean, and the nest dark.

206. *Roost Support.* A simple wire roost support is shown. To make it you will need a piece of stout galvanized wire, pieces of heavy galvanized iron for the wall supports and eight round-headed screws. The lengths of the different sides of the support will be regulated by the size of the roost, but the length of the top A should be 2 in. The roost should come within $\frac{1}{2}$ in. of the wall,

and a wire nail driven into the end as shown will keep it in place and prevent its moving endways. If you make a new roost have the rounded top on which the fowls rest 2 in. wide, and the depth 3 in. The roost should be planed on four sides and frequently painted with liquid lice killer, or kerosene and coal tar, to prevent lice or mites living on it.

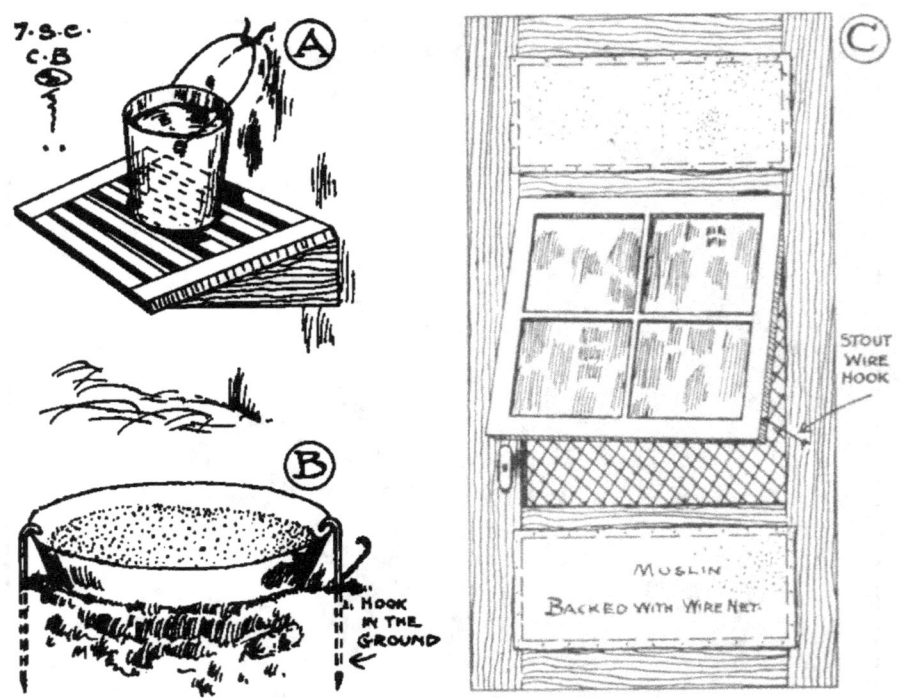

MR. BATCHELDER'S APPLIANCES AND FRESH-AIR DOOR

(A) rack for water pail. (B) arrangement for preventing food or water pans being overturned. (C) fresh-air or ventilating door. The common muslin used.

207. *Wrinkles.* Mr. Batchelder illustrates two useful appliances and a fresh-air door. The rack for the water pail is held in place by brackets, and is 12 in. above the floor. A common galvanized iron pail is used and there is a hook in the wall for the handle to catch on. The water and shelf will keep clean and the pail cannot be tipped over.

This poultryman also shows a simple device to prevent water and feed pails being upset. Two wires, about the diameter of a lead pencil and long enough to form hooks on the upper end and extend into the ground a few inches at the lower end, are placed as shown in the illustration. This effectually prevents the chicks or fowls from tipping the water or feed pans over, and will thereby save many useless steps — to say nothing of the waste.

208. *A Fresh-Air Door.* Mr. Batchelder also uses a fresh-air or ventilating door. This may be fitted to any poultry house, but preferably to small colony houses that depend upon one window

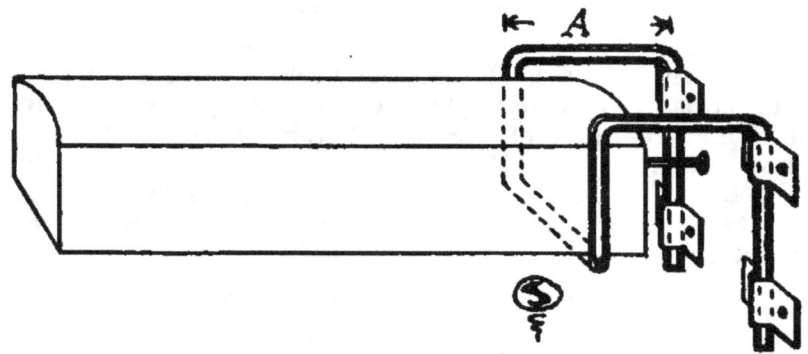

A SIMPLE WIRE ROOST SUPPORT

for light, and have but one door, and that at the center of the sunny side. It is so simple that its construction is readily understood. Common unbleached muslin over 1-in. mesh wire netting covers the ventilating openings. "I have given this door a good trial the year around," said Mr. Batchelder, "and I find that it works perfectly on a shed-roof, 7 x 9-ft. colony house facing the south. No other door, window or ventilator is necessary."

To the Reader — Have you built and used a house or any poultry appliance that is "just what the doctor ordered?" If you have, will you not send us a description of it with a rough pencil sketch if you can, so we can include it in the Fourth edition of this work? We hope to improve on this edition and keep the designs right up to date — and all practical. — THE EDITOR.

www.ingramcontent.com/pod-product-compliance
Lightning Source LLC
Chambersburg PA
CBHW080832220526
45467CB00008B/2261